Impacts of Agrochemicals: Environmental Fate, Ecotoxicology, Risk Assessment, and Remediation

Impacts of Agrochemicals: Environmental Fate, Ecotoxicology, Risk Assessment, and Remediation

Guest Editors

Eszter Takács
Szandra Klátyik
Mária Mörtl
András Székács

MDPI

Basel • Beijing • Wuhan • Barcelona • Belgrade • Novi Sad • Cluj • Manchester

Guest Editors

Eszter Takács
Institute of Environmental Sciences
Hungarian University of Agriculture and Life Sciences
Gödöllő
Hungary

Szandra Klátyik
Institute of Environmental Sciences
Hungarian University of Agriculture and Life Sciences
Gödöllő
Hungary

Mária Mörtl
Institute of Environmental Sciences
Hungarian University of Agriculture and Life Sciences
Gödöllő
Hungary

András Székács
Institute of Environmental Sciences
Hungarian University of Agriculture and Life Sciences
Gödöllő
Hungary

Editorial Office
MDPI AG
Grosspeteranlage 5
4052 Basel, Switzerland

This is a reprint of the Special Issue, published open access by the journal *Toxics* (ISSN 2305-6304), freely accessible at: https://www.mdpi.com/journal/toxics/special_issues/pesticides_ecotoxicology_risk.

For citation purposes, cite each article independently as indicated on the article page online and as indicated below:

Lastname, A.A.; Lastname, B.B. Article Title. *Journal Name* **Year**, *Volume Number*, Page Range.

ISBN 978-3-7258-5933-7 (Hbk)
ISBN 978-3-7258-5934-4 (PDF)
https://doi.org/10.3390/books978-3-7258-5934-4

Contents

About the Editors

Eszter Takács

Eszter Takács is a senior research fellow and the head of the Department of Ecotoxicology within the Agro-Environmental Research Center at the Institute of Environmental Sciences, Hungarian University of Agriculture and Life Sciences in Gödöllő, Hungary. Her research activities include the development and application of enzyme-linked immunosorbent assay (ELISA) methods for the quantification of toxin content in genetically modified plants, in plant and animal tissues, and for monitoring the environmental fate of organic micropollutants, as well as analytical characterization (calibration and internal quality control) of commercially available ELISA systems and their applicability in plant and animal tissues and ecotoxicity and cytotoxicity studies of active substances and additives in herbicide formulations. She is a member of the Hungarian Society of Ecotoxicology. Her number of peer-reviewed publications: 41, h-index: 15, and number of independent citations: 494.

Szandra Klátyik

Szandra Klátyik, PhD in biology, is a research fellow at the Agro-Environmental Research Centre within the Institute of Environmental Sciences at the Hungarian University of Agriculture and Life Sciences. Her research work is primarily related to the ecotoxicological assessment of agricultural pollutants (e.g., toxins produced by first-generation genetically modified plants, veterinary medicines, and pesticide formulations) on non-target plant and animal organisms using various standardized and self-developed test systems at the individual and community levels. Additional research areas cover the monitoring of the environmental fate of pollutants and the application of enzyme activity testing methods and biomarkers in toxicity assessment. She has participated in several national and international projects. She is the Executive Vice President of the Hungarian Society of Ecotoxicology. Her number of peer-reviewed publications: 27, h-index: 11, and number of independent citations: 313.

Mária Mörtl

Mária Mörtl, PhD in chemistry, is a senior research fellow at the Hungarian University of Agriculture and Life Sciences (MATE) (Institute of Environmental Sciences, the Agro-Environmental Research Centre). Earlier, she worked for twenty years at Eötvös Loránd University (ELTE) in the field of organosilicon chemistry, and later, she engaged in research on environmental problems in different research institutes. Her main research fields are currently related to instrumental analysis (chromatography, mass spectroscopy, derivatization) and environmental fate of pollutants (monitoring, uptake by plants, dissipation). The contaminants studied are often of agricultural origin, including mycotoxins, pesticides, surfactants, anthropogenic contaminants, etc.). His number of peer-reviewed publications: 52, cumulated impact factor: 137.2, and h-index: 15.

András Székács

András Székács, PhD in chemistry, Doctor of the Hungarian Academy of Sciences, is a professor at the Hungarian University of Agriculture and Life Sciences (MATE), Deputy Director of the Institute of Environmental Sciences, and Head of the Agro-Environmental Research Centre. He is also an Honorary Professor at the Budapest University of Technology and Economics (BME). His main fields of research are chemical and environmental safety of agricultural and chemical technologies (including mycotoxins, pesticides, genetically modified crops, surfactants, anthropogenic contaminants, etc.), as well as instrumental and immunoanalysis of organic micropollutants. He served as a member and Vice Chair of the Management Board of the European Food Safety Authority (2014–2022) and as a member and Vice Chair of the Scientific Advisory Body of the OECD Co-operative Research Programme (2015–2022). His number of peer-reviewed publications: 239, cumulated impact factor: 474.6, and h-index: 32.

toxics

MDPI

Article

Hormesis, the Individual and Combined Phytotoxicity of the Components of Glyphosate-Based Formulations on Algal Growth and Photosynthetic Activity

Szandra Klátyik [1], Eszter Takács [1], Attila Barócsi [2], Sándor Lenk [2], László Kocsányi [2], Béla Darvas [3] and András Székács [1,*]

[1] Agro-Environmental Research Centre, Institute of Environmental Sciences, Hungarian University of Agriculture and Life Sciences, H-2100 Gödöllő, Hungary; klatyik.szandra@uni-mate.hu (S.K.); takacs.eszter84@uni-mate.hu (E.T.)
[2] Department of Atomic Physics, Institute of Physics, Budapest University of Technology and Economics, H-1111 Budapest, Hungary; barocsi.attila@ttk.bme.hu (A.B.); lenk.sandor@ttk.bme.hu (S.L.); kocsanyi.laszlo@ttk.bme.hu (L.K.)
[3] Hungarian Society of Ecotoxicology, H-1022 Budapest, Hungary; mott@bdarvas.hu
* Correspondence: szekacs.andras@uni-mate.hu

Abstract: The occurrence of the market-leading glyphosate active ingredient in surface waters is a globally observed phenomenon. Although co-formulants in pesticide formulations were considered inactive components from the aspects of the required main biological effect of the pesticide, several studies have proven the high individual toxicity of formulating agents, as well as the enhanced combined toxicity of the active ingredients and other components. Since the majority of active ingredients are present in the form of chemical mixtures in our environment, the possible combined toxicity between active ingredients and co-formulants is particularly important. To assess the individual and combined phytotoxicity of the components, glyphosate was tested in the form of pure active ingredient (glyphosate isopropylammonium salt) and herbicide formulations (Roundup Classic and Medallon Premium) formulated with a mixture of polyethoxylated tallow amines (POEA) or alkyl polyglucosides (APG), respectively. The order of acute toxicity was as follows for Roundup Classic: glyphosate < herbicide formulation < POEA. However, the following order was demonstrated for Medallon Premium: herbicide formulation < glyphosate < APG. Increased photosynthetic activity was detected after the exposure to the formulation (1.5–5.8 mg glyphosate/L and 0.5–2.2 mg POEA/L) and its components individually (glyphosate: 13–27.2 mg/L, POEA: 0.6–4.8 mg/L), which indicates hormetic effects. However, decreased photosynthetic activity was detected at higher concentrations of POEA (19.2 mg/L) and Roundup Classic (11.6–50.6 mg glyphosate/L). Differences were demonstrated in the sensitivity of the selected algae species and, in addition to the individual and combined toxicity of the components presented in the glyphosate-based herbicides. Both of the observed inhibitory and stimulating effects can adversely affect the aquatic ecosystems and water quality of surface waters.

Keywords: glyphosate; co-formulants; POEA; APG; algae; phytotoxicity; photosynthetic activity; hormesis; growth inhibition; combined toxicity

1. Introduction

The majority of various pesticide formulations have significant direct or indirect detrimental effects on the environment, particularly in surface waters due to their leaching, surface run-off from treated areas, drifting, foliar spray, and unintended overspray [1–3]. Non-selective glyphosate-based herbicides (GBHs) are no exemption from this trend [4,5]. Originally, these herbicides were exclusively applied for pre-emergence weed control. However, the introduction of glyphosate-tolerant genetically modified crops (not authorized for cultivation in the European Union) and the adoption of pre-harvest desiccation practices in agriculture resulted in a substantial increase in the use of glyphosate-based formulations [6–8]. However, the approval of the active substance glyphosate has been renewed according to the current legislation subject to the specified conditions and restrictions. Based on the Commission Implementing Regulation (EU) 2023/2660, pre-harvest use of GBHs as desiccants to control the time of harvest or optimize threshing is not authorized [9,10].

Globally, more than 2000 commercial GBHs are used for chemical plant protection against weeds. Different salts of glyphosate (e.g., glyphosate isopropylammonium salt, glyphosate diammonium salt, or glyphosate trimethylsulfonium salt) are used as active ingredients in various GBHs to enhance the solubility of the parent compound, glyphosate [11,12]. In addition to the active ingredient, various co-formulants are also included in GBHs. The primary function of these co-formulants is to facilitate the effectiveness and bioavailability of the formulation by increasing the solubility, adsorption, and absorption of the active ingredient [13]. For example, POEA (a mixture of polyethoxylated tallow amines) as a formulating agent in GBHs enhances the penetration of glyphosate into the plant cell [14]. Various co-formulants presented in commercial pesticides were considered inactive components with regard to the required main biological effect of the formulation. However, numerous studies have indicated the high individual toxicity of co-formulants and the enhanced combined toxicity of the active ingredients and co-formulants in various commercial pesticide formulations compared to the individual toxicity of active ingredients [15–17]. Therefore, the use of POEA in GBHs has been banned in the EU due to the incriminating scientific evidence [18].

As a result of excessive global use, glyphosate has become a ubiquitous contaminant in aquatic ecosystems [19,20]. The appearance and concentration of glyphosate in the different environmental elements (e.g., soil, ground and surface waters) are highly influenced by several abiotic (e.g., hydrological conditions, pH, suspended materials), biotic (e.g., activity and composition of the microbial community), and climatic factors (e.g., rainfall frequency and intensity) [21–23], in addition to the condition of pesticide treatments (e.g., frequency and timing of the treatment) [22,24]. In the past, glyphosate was not included in standard pesticide monitoring programs. Thus, the environmental concentration of glyphosate and its metabolites were underestimated particularly in regions where pre-harvest desiccation practices are widespread or the cultivation of glyphosate-tolerant genetically modified crops occurs extensively. The primary metabolite of glyphosate, aminomethylphosphonic acid (AMPA), is more mobile in water than the parent compound [25,26] and is frequently detected in various environmental elements, including groundwater and surface waters [26–31]. However, it is important to note that the appearance of AMPA in environmental matrices (e.g., groundwater, influents, or sewage sludge) is not exclusively a result of glyphosate metabolism, as it can also originate from phosphonate detergents used in different softeners and cleaning agents [31,32]. According to the U.S. Geological Survey, the presence of glyphosate and/or AMPA was identified in 59% of the analyzed surface waters [33].

The level of glyphosate contamination can reach up to 5.2 mg/L in surface water, although mainly in streams near the treated agricultural fields and especially after heavy rains [27,34,35]. However, high variability can be observed in the detected glyphosate residue levels in various surface water samples [27]. In surface waters collected in Argentina, the average concentrations of glyphosate and AMPA were in the ranges of 17.5–35.2 and 0.6–2.1 μg/L, respectively [36]. However, maximum concentrations were up to 0.258 and 5.87 mg/L, respectively, in the analyzed groundwater and surface water samples [37]. Based on European monitoring programs, the level of glyphosate contamination in surface waters within the EU seems to be relatively lower, with typical glyphosate concentrations ranging between 0.05 and 0.85 μg/L, but residues are consistently detectable [27]. In water samples collected from Hungary, Switzerland, and Italy, the detected glyphosate contamination ranged from 0.035 to 96 μg/L [5,27,35,38–40].

Different GBHs manufactured with various co-formulants show different environmental behaviors (e.g., different half-lives and mobility in soil and water). After the pesticide treatments, the active ingredients and the co-formulants rapidly become separated in most cases. The half-life (DT_{50}) of glyphosate in water varies from a few to 91 days [41]. Furthermore, the photo- and biodegradation of the active ingredient also occurred in surface waters [41,42], although limited information is available about the half-lives and the environmental fate of the co-formulants [20,43]. In general, most studies focus on the possible toxic effects on various aquatic organisms and the analytical possibilities of the qualitative and quantitative determination of the co-formulants including POEA and APGs. However, the presence of the GBH co-formulant POEA has been observed extensively in soils collected from agricultural fields of the mid-western states in the USA, where the cultivation of genetically modified glyphosate-tolerant crops is concentrated [44]. Additionally, studies have demonstrated the persistence of POEA in soil along with glyphosate and AMPA [44–46] and possible access to natural waterways [20,43,45]. Due to their low environmental impacts and biodegradability [47], APGs are commonly used as additives in pesticide formulations, personal skin products, and drugs [48,49]. The environmental concentration of co-formulants is generally not monitored [45]; therefore, the exact concentration of POEA and APGs in surface waters is not known. However, the presence of co-formulants such as POEA and APGs in the environment has been demonstrated (e.g., soil, sediment, wastewater) [43,49]. Typically, glyphosate and the co-formulants presented in the GBHs coexist in environmental matrices (e.g., soil and waters) and such co-exposure can affect various non-target aquatic organisms. The aquatic organisms and communities are highly exposed to water pollution [50], as their contact with xenobiotics in water is unavoidable. Recently, the possible combined toxic effects between active ingredients and co-formulants on the environment and non-target organisms are poorly understood.

The toxic effects of glyphosate and its formulations have been studied in numerous aquatic organisms, such as various algae species [51,52], crustaceans (e.g., *Daphnia magna*) [53], mollusks [54], fish [55], and amphibians [56]. Based on the results of ecotoxicological testing performed on a wide range of aquatic plant and animal organisms, the damage to different physiological and behavioral functions was demonstrated [20]. In aquatic ecosystems, algal communities constitute the primary producer level and the majority of biomass, playing a key role in the oxygen cycle of water and the atmosphere. They also have essential roles in aquatic food webs and nutrient transport processes [57,58]. In addition, some species (e.g., *Ankistrodesmus fulcatus*) can participate in the breakdown of organic pollutants and toxic compounds (e.g., tributyltin) [59]. However, the massive proliferation of certain cyanobacterial (e.g., *Anabaena flos-aquae, Microcystis aeruginosa*) [60] and green algae species (e.g., *Pleodorina indica*) can lead to deterioration of the water quality of surface waters [61]. The determination of the effects of herbicides used for chemical plant protection on algal species is crucial for the toxicological assessment of herbicide formulations. Various algal species are widely used for environmental biological monitoring [62] and bioremediation activities [63]. The different effects of glyphosate, its metabolite, co-formulants, and/or commercial herbicide formulations on green algae and cyanobacterial species are summarized in Table 1 partially based on our previous review about aquatic ecotoxicity of glyphosate, its formulations, and co-formulants [20].

Based on the results of ecotoxicological studies, the inhibitory effects of glyphosate and its formulation on various green algal (e.g., *Chlorella vulgaris, Scenedesmus incrassatulus, Pseudokirchneriella subcapitata*) [64,65] and cyanobacterial (e.g., *M. aeruginosa*) species were observed [66]. However, stimulated growth was observed at lower test concentrations after exposure to glyphosate and glyphosate-based formulations [64,65,67].

In addition, altered cell morphology, disrupted ultrastructure (e.g., damaged thylakoids and mitochondria) as well as altered biochemical and physiological parameters (e.g., antioxidant activity, lipid peroxidation) were also demonstrated in algae [52,68,69]. Additionally, differences were observed in the sensitivity of the investigated aquatic organisms, even with similar lifestyles, habitats, or identical taxa [70–72]. For example, the determined 72 h EC_{50}

values are in the range of 24.7–166 mg/L [15,20,73] for *P. subcapitata*, while for *Desmodesmus subspicatus* higher values were calculated (72.9–166 mg/L) [41,52,74–76] during the ecotoxicological testing on the effects of glyphosate. Moreover, potential adverse effects of glyphosate and GBHs were indicated also on freshwater periphyton [77–79]. Moreover, the increased toxicity of Roundup was demonstrated on cyanobacterial and green algal species (*M. aeruginosa, Nitella microcarpa var. wrightii*) in the presence of POEA [80]. The 72 h EC_{50} values for POEA in *P. subcapitata* ranged from 0.2 to 4.9 mg/L [15,81,82]. The negligible aquatic toxicity of APGs was demonstrated on *P. subcapitata* [83], but the toxicity of APGs highly depends on the length of the of the carbon chain [84,85].

In addition to growth inhibition, photosynthetic activity is a commonly used endpoint during the assessment of phytotoxic effects. The measurement of photosynthetic parameters ensures a non-invasive and rapid indication of harmful effects. A widely used method for the measurement of photosynthetic activity is the detection of induced chlorophyll-a fluorescence [86]. Recently, the measurement of photosynthetic activity is widely used in research on stress effects on plant organisms, and for characterizing the physiological state of plants [87–89]. In addition to herbicide active ingredients that directly inhibit photosynthesis (e.g., atrazine), additional active ingredients, including glyphosate, can also impact photosynthetic and respiratory processes by influencing various metabolic pathways [90,91]. The adverse effects of glyphosate on photosynthetic processes can be explained by the direct or indirect inhibition of plastoquinone biosynthesis [92,93]. Furthermore, the reduction in chlorophyll concentration [94] directly affects the rate of electron transport in the chloroplast [91]. Reactive oxygen species generated in mitochondria can further affect photosynthesis by inhibiting the respiratory electron transport chain as a result of glyphosate exposure. The generated free radicals leave mitochondria and enter the chloroplast, where they cause oxidative damage to the photosynthetic apparatus and reduce the activity of photosynthesis [94]. The phytotoxic effects of glyphosate and its herbicide formulation on photosynthetic activity have been studied on phytoplankton species [95,96]. The observed effects indicated damage to the photochemical efficiency of the PS II photochemical system [95]. However, increased growth, chlorophyll-a content, and photosynthetic activity were observed at lower concentrations [90].

The aim of this study was to assess the individual and combined acute phytotoxicity of the components of glyphosate-based formulations. During the comparison of toxic effects, glyphosate was tested in the form of pure active ingredient (glyphosate isopropylammonium (IPA) salt) and preparations (Roundup Classic and Medallon Premium) formulated with a mixture of polyethoxylated tallow amines (POEA) or alkyl polyglucosides (APG), respectively. In addition, the individual toxicity of the formulating agents (POEA and APG) was also investigated. During our study, standard algal growth inhibition assays were performed on different green algae species (*D. subspicatus, P. subcapitata, Scenedesmus obtusiusculus*) and a cyanobacteria (*A. flos-aquae*). Based on the results, the differences in the sensitivity of various algal species were also compared. In addition, we investigated the possible effects on the photosynthetic activity of *P. subcapitata* algae cells exposed to the components of Roundup Classic individually and in combination.

Table 1. Effects of glyphosate, its metabolite, co-formulants, and/or commercial herbicide formulations on green algae and cyanobacterial species.

Algae Species	Tested Substances	Test Concentrations	Test Period	Tested Parameters	Main Results	Reference
P. subcapitata	technical-grade glyphosate (GLY) acid, GLY-IPA [a], Roundup, POEA [b]	dilution series	96 h [c]	growth inhibition	96 h IC_{50} [d] = 3.92 mg a.e. [e] /L (POEA), 5.81 mg a.e./L (Roundup), 24.7 mg a.e./L (GLY acid), 41.0 mg a.e./L (GLY-IPA)	[15]
P. subcapitata	Roundup	4.7–60 mg/L	96 h	growth inhibition	96 h EC_{50} [f] = 15.60 mg/L, damaged cell ultrastructure	[52]
C. vulgaris	GLY, AMPA [g]	0.05–50 mg/L, individual and co-exposures	7 d [h]	growth inhibition, pigment content, antioxidant activity	stimulated growth ($\leq$ 0.5 mg/L), growth inhibition ($\geq$ 5 mg/L), inhibitory effect ($\geq$ 5 mg/L GLY and AMPA), altered pigment levels, increased antioxidant activity	[64]
cyanobacteria, *Chlorophycean* microalgae	GBH [i] (Faena)	1–100 mg/L	96 h	growth inhibition, antioxidant enzymes	IC_{50} = 1.022–2.702 mg/L, affected antioxidant enzyme activity ($\geq$ 0.74 mg/L)	[65]
M. aeruginosa	GLY	1–10 mg/L	9 d, enzyme assays: 24–48 h	growth inhibition, chl-a [j] content, antioxidant activity, cell apoptosis	reduced growth and chl-a content, increased antioxidant activity (1–2 mg/L), induced apoptosis	[66]
cyanobacterial strains	GLY	8.5–33.8 mg/L	15 d	growth inhibition, phosphate and phosphonate levels	species- and dose-dependent stimulatory effects, decreased phosphonate levels, concentration-dependent phosphate uptake	[67]
S. vacuolatus	GBH (Glifosato Atanor) with 2.5% of the surfactant (alkyl aryl polyglycol ether)	0–8 mg GLY/l	96 h	growth, morphology, oxidative stress parameters	96 h IC_{50} = 4.9 mg/L, metabolic and morphological changes ($\geq$ 4 mg/L), oxidative damage ($\geq$ 6 mg/L)	[68]
cyanobacterial species	pesticide adjuvants	dilution series	96 h	growth inhibition	substance- and species-specific effects	[71]
N. microcarpa var. *wrightii*	technical-grade GLY, GBH (Roundup), AMPA	GLY, Roundup: 0.28, 3.5, 6 mg/L; AMPA: 0.03 mg/L	7 d	photosynthetic rate, dark respiration rate, chl-a	higher toxicity of Roundup, stimulatory effect of AMPA	[80]
P. subcapitata	POEA	dilution series	96 h	growth inhibition	96 h EC_{50} = 4.1–4.9 mg/L	[81]
P. subcapitata, *C. vulgaris*, *Oophila sp*	MON 0818	dilution series	96 h	growth inhibition	96 h EC_{50} = 0.21–1.61 mg/L	[82]
P. subcapitata	APG [k]	dilution series	72 h	growth inhibition	negligible aquatic toxicity	[83]
P. subcapitata	APG	dilution series	72 h	growth inhibition	toxicity affected by the length of the carbon chain	[84]
green microalgae species	APG	0.26–6.8 mg/L	72 h	growth inhibition	72 h EC_{50} = 0.32–2.7 mg/L	[85]
M. aeruginosa	GLY, Roundup	0.06–29.6 µg/L	21 d	cell number, chl-a, APA [l] activity	increased cell number and chl-a, inhibition (> 5.92 µg/L), GLY increased photosynthesis, concentration-dependent APA activity	[90]
freshwater microalgae	GLY	maximum tested concentration: 5.07 g/L	80 min	chl-a fluorescence, cell viability	concentration-specific effect on maximum quantum yield of PSII [m] (< 0.17 mg/L)	[95]
microalgal and cyanobacterial species	Factor 540R	10–1000 µg/L	48 h	growth inhibition, photosynthetic parameters	48 h EC_{50} = 406–724 µg/L, modified photosynthetic response ($\geq$ 10 µg/L)	[96]

[a] glyphosate isopropylammonium salt; [b] mixture of polyethoxylated tallow amines; [c] hour; [d] half-maximal inhibitory concentration; [e] acid equivalent; [f] 50% effective concentration; [g] aminomethylphosphonic acid; [h] day; [i] GBH: glyphosate-based herbicide; [j] chlorophyll-a; [k] alkyl polyglycoside; [l] alkaline phosphatase activity; [m] photosystem II.

2. Materials and Methods

2.1. Standard and Reagents

Glyphosate IPA salt and the mixture of polyethoxylated tallow amines (POEA, under the tradename: Emulson AG GPE 3SS) were received from Lamberti SpA (Albizzate, Italy). The glyphosate-based Roundup Classic (Monsanto Europe S.A./N.V.) [97] and Medallon Premium (Syngenta) [98], in addition to the alkyl polyglucosides (APG, under the tradename: Plantapon LGC) were purchased from a public commercial source. The main chemical properties of the investigated herbicide active ingredient, formulations, and surfactants (POEA and APG) used in the investigated formulations can be found in Table 2. Based on the Material Safety Data Sheet (MSDS), Roundup Classic contains 41.5% glyphosate IPA salt and 15.5% POEA. In addition, Medallon Premium consists of 34% glyphosate diammonium salt and 10–20% APG. However, the selected formulations contain different salts of glyphosate, and the indicated concentrations of the active ingredient correspond to 360 g/L glyphosate acid concentration for both preparations. During the ecotoxicological testing, glyphosate was tested only in the form of glyphosate IPA salt. In the tested concentration ranges, the water solubility is not limited for any forms of glyphosate active ingredient, and in water, the salts of glyphosate quickly dissociate into ions that are also found in nutrient solutions and buffer solutions used in ecotoxicological studies.

Table 2. Composition and chemical characteristics of the investigated chemical substances.

Active ingredient (AI)					
Chemical Name	CAS No.[1]	Concentration of the AI	Physical Appearance	Chemical Structure	
glyphosate isopropylammonium (IPA) salt	38641-94-0	62% (486 g/L glyphosate acid)	water-soluble emulsion	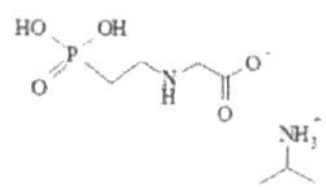	
Glyphosate-based formulations					
Product name	AI	Concentration of the AI	Co-formulants	Concentration of the co-formulants	Type of formulation
Roundup Classic	glyphosate IPA salt	41.5% (360 g/L glyphosate acid)	mixture of polyethoxylated tallow amines (POEA)	15.5%	liquid water-soluble concentrate
Medallon Premium	glyphosate diammonium salt(CAS 69254-40-6)	34% (360 g/L glyphosate acid)	alkyl polyglucosides (APG)	10–20%	liquid water-soluble concentrate
Co-formulants					
Product name	Co-formulant	Concentration of the co-formulant	Additives	Type of formulation	Chemical structure
Emulson AG GPE 3SS	POEA (CAS 61791-26-2)	100%	–	water-soluble emulsion	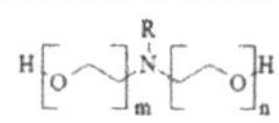
Plantapon LGC	APG (Na-lauryl glucose carboxylate CAS 383178-66-3 + lauryl glucoside CAS 110615-47-9)	28.5–34.0%	water: 66–71.5%	water-soluble emulsion	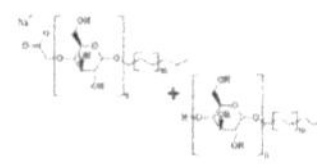

[1] Chemical Abstracts Service (CAS) Registry Number.

2.2. Selected Algae Monocultures

The selected algae species were obtained from public collections. The green algae *Pseudokirchneriella subcapitata*, Korshikov (NIVA-CHL1, previous name: *Selenastrum capricornutum*, current name: *Raphidocelis subcapitata*) was obtained from the alga collection of the Norwegian Institute for Water Research. The additional *Desmodesmus subspicatus*, Hegewald & Schmidt (CCAP 276/20) and *Scenedesmus bijugus var. obtusiusculus*, Schmidt (CCAP 276/25) green algae species, as well as the investigated filamentous cyanobacteria *Anabaena flos-aquae* (CCAP 1403/13D, current name: *Dolichospermum flos-aquae*), were derived from the Scottish Culture Collection of Algae and Protozoa. The batch culture of green algae species and the selected cyanobacteria were maintained in Zehnder-8 (pH = 6–7) [99] and Allen (pH = 6–7) [100] media, respectively. Fresh media were added to the algae cultures every two weeks, and they were maintained at 20 ± 2 °C and illuminated in a 14:10 light/dark period with the use of cool-white fluorescence tubes (15 μmol/m^2/s). The sensitivity of the algae cultures was verified with the use of the reference substance (potassium dichromate, $K_2Cr_2O_7$) before testing and was proven to be acceptable (72 h EC_{50} = 1.0 ± 0.1 mg/L) within the appropriate ranges (0.8 ± 0.1 mg/L for *D. subspicatus*; 1.2 ± 0.3 mg/L for *P. subcapitata*) based on the relevant standard protocol [101].

The selected green algal and cyanobacterial strains are sensitive to changes in water quality, so they serve as excellent test organisms for the investigation of the toxic effects of aquatic pollutants. *P. subcapitata* and *D. subspicatus* are also considered reference species recommended by the related OECD guideline [102]. Additionally, the selected strains can be easily maintained under laboratory conditions, and are characterized by a fast reproduction and life cycle [103,104]. Based on the scientific literature, significant differences can be observed in the sensitivity of algal species to certain pollutants even within the same taxa [71]. To investigate and compare the sensitivity of taxonomically close and distant species to the applied treatments, three common representatives of freshwater green algae (Phylum: *Chlorophyta*) were selected. Two of the selected green algae species (*D. subspicatus* and *S. obtusiusculus*) belong to the same taxonomic family (*Scenedesmaceae*), thus the sensitivity can be compared also in the case of taxonomically very close species. With the use of *A. flos-aquae* representing a species of cyanobacteria known for its ability to form harmful algal blooms [60], the differences in the sensitivity to the effects of glyphosate can be evaluated for eukaryotic green algae cells and a prokaryotic cyanobacterium as well.

2.3. Algal Growth Inhibition Tests

The individual and combined phytotoxic effects of the components of the tested GBHs were evaluated in algal growth inhibition tests based on the OECD 201 guideline [102]. Growth inhibition tests were performed on three unicellular green algae species (*P. subcapitata, D. subspicatus, S. obtusiusculus*) and, in the case of the active ingredient glyphosate, the tests were also performed on the selected filamentous cyanobacteria (*A. flos-aquae*). The duration of the test was 72 h. During the tests, continuous and uniform cool-white illumination (104.9–14.9 μE/m^2/s), optimal pH of algal media (pH = 6–7 for Zehnder-8 and Allen media, as well), controlled temperature (22 ± 2 °C) and stirring (continuous, 100 rpm) were ensured in a shaking incubator (Witeg WIS-10RL, Wertheim, Germany) [102]. The tested compounds were serially diluted, and five concentrations of the substance along with the control were investigated in three repetitions at each level. Each test was repeated three times for each investigated compound. The initial number of algae cells was 10^5 cells/mL in the tested and control groups with the fulfillment of the conditions for exponential growth during the entire exposure time. During the growth inhibition assays, the algal cell density was determined daily in the control group to monitor the required specific reproduction rate. The concentration ranges used in the algal growth inhibition tests were as follows for the different species: (1) *P. subcapitata*: glyphosate IPA salt: 22–352 mg/L, Roundup Classic: 3.5–56 mg/L, Medallon Premium: 45–720 mg/L, POEA: 0.5–8 mg/L, APG: 6.5–104 mg/L; (2) *D. subspicatus*: glyphosate IPA salt: 22–352 mg/L, Roundup Classic: 7–112 mg/L, Medallon Premium: 95–1520 mg/L, POEA: 0.8–13 mg/L, APG: 10–160 mg/L;

(3) *S. obtusiusculus*: glyphosate IPA salt: 22–352 mg/L, Roundup Classic: 15–240 mg/L, Medallon Premium: 125–2000 mg/L, POEA: 1.5–24 mg/L, APG: 30–480 mg/L; (4) *A. flos-aquae*: glyphosate IPA salt: 4.5–36 mg/L.

At the end of the experiments, we determined the amount of algal biomass in each control and treated group. Algal biomass was characterized by the measurement of optical density and chlorophyll-a content. The optical density of green algae cells was determined at a wavelength of 750 nm using a spectrophotometer (UV/VIS Camspec single beam M330, Camspec, Crawley, UK), also in three repetitions for each sample [101]. In addition to the measurement of optical density, the potential toxic effects were also evaluated based on the chlorophyll-a content of the samples in the tests performed on green algae species exposed to the POEA-formulated herbicide and its components. Due to the filamentous structure of *A. flos-aquae*, more reliable results were obtained with the measurement of the chlorophyll-a content. The correlation between the two test methods proved to be very high in the case of green algae species ($R^2 > 0.998$). After the extraction process, the chlorophyll-a content of the samples was also determined using a spectrophotometric method in the three replicates [105]. In the performed tests, the coefficient of variation for section-by-section specific growth rates (days 0–1, 1–2, and 2–3) remained below 35% in the control groups. During the entire duration of the tests, the coefficient of variation of the specific growth rates did not exceed 7% in the parallel control cultures of *P. subcapitata* and *D. subspicatus*. In addition, more than 16-fold growth was detected in the control groups, thus the tests can be considered valid.

During the testing of individual and combined effects on algal growth, algae cells were exposed to glyphosate IPA salt and the tested surfactants (POEA and APG) individually and in the form of formulated herbicides. The individual and combined toxicity of the tested substances was evaluated by the determined 72 h EC_{50} values. The 72 h EC_{50} values were calculated based on the measured optical densities and chlorophyll-a contents as well. During the comparative study of the individual and combined toxicity, the 72 h EC_{50} values determined for the tested GBH formulations were corrected with the nominal content of the active ingredient glyphosate and the surfactant as well, based on the MSDS (Table 2).

2.4. Photosynthetic Activity Tests

The individual and combined effects of the components of Roundup Classic were assessed on the photosynthetic activity of *P. subcapitata* green algae. The photosynthetic activity was determined in the samples derived from the algal growth inhibition tests after the 72 h exposure. The measurements were carried out with a portable FluoroMeter Module (FMM) device based on the detection of laser-induced chlorophyll-a fluorescence [106]. The measuring principle of the instrument is based on the "Kautsky effect" [86]. Under dark conditions, the photochemical process of photosynthesis in plant cells temporarily ceases, and upon sudden high-intensity stimulation, typically by laser excitation, the chlorophyll molecules in the cells immediately begin to absorb light. However, the optimal conditions for photosynthesis develop more slowly, so only a small fraction of the energy of the light absorbed at the beginning is used in the process of photosynthesis. The excess light energy is re-emitted by the cells in the form of fluorescent light. After a few minutes, as the photosynthetic process resumes, the plant cell utilizes the absorbed light with higher efficiency, causing the intensity of fluorescent radiation to gradually decrease, stabilizing at a lower value [86,107].

During the 96-well microplate-based assay, the photosynthetic parameters were measured after a 10 min dark adaptation with the use of a special sample holder. After the dark acclimatization, the samples were excited with a laser diode (10 mW) at the wavelength of 635 nm. After excitation, the duration of the measurement was 5 min. During the measurements, the intensity of the fluorescent light emitted by the sample was detected at wavelengths of 690 nm and 735 nm [106,107]. The measurements were performed in triplicates. Photosynthetic activity of the control and treated groups was characterized by the observed ratio of fluorescence decrease (Rfd*) and the proxy of quantum efficiency of the algae photosystem PSII (Fv*/Fp). Here, Fp means the peak fluorescence value derived from the fluorescence induction curve using the FMM module, while Fv* represents the variable fluorescence in terms of Fp. Essentially, the Fv/Fm parameter describes the impact of plant stress on photosystem II in a dark-acclimated state, where Fm is the maximum chlorophyll fluorescence under a saturating radiation pulse in such conditions [108,109]. As the maximum actinic level available with the FMM will not saturate PSII, Fp is used to distinguish it from Fm, which represents the maximum fluorescence value during continuous excitation under full saturation [88,108]. Rfd* corresponds to Fd/Fs, where Fs is the observed steady-state fluorescence and Fd indicates the fluorescence reduction from Fp to Fs [106,108]. Detailed explanations of different fluorescence parameters are summarized in Table 3. The effects on photosynthetic activity were compared based on the values detected at the wavelength of 690 nm [88].

Table 3. The main fluorescence parameters and quantities determined by FMM [106].

Fluorescence Parameter	Definition	Interpretation
Fo	observed	Non-variable (original) fluorescence intensity
Fp	observed	Peak fluorescence intensity, maximum fluorescence at a non-saturating light pulse
Fv*	Fp–Fo	Variable fluorescence in terms of Fp
Fv*/Fp	Fv*/Fp	Proxy of quantum efficiency of photosystem II
Fs	observed	Steady-state (terminal) fluorescence
Fd	Fp–Fs	Fluorescence decrease in terms of Fp
Rfd*	Fd/Fs	Fluorescence decrease ratio

2.5. Statistical Analysis

Based on the results of algal growth inhibition tests, 72 h EC_{50} values for both measured parameters (optical density and chlorophyll-a content) were determined using the ToxRat Pro 3.0 statistical software (ToxRat Solutions Gmbh, Alsdorf, Germany). The additional statistical analyses were performed with the use of the R Statistical program 4.2.1. (R Development Core Team, Vienna, Austria). The effects of the individual and combined exposures, in addition to the differences between the determined 72 h EC_{50} values, and the detected parameters of photosynthetic activity (Fv*/Fp and Rfd*) were evaluated with the use of general linear models. Before the statistical analysis, the normality of the data and the homogeneity of variance were checked by Shapiro–Wilk and Levene's or Bartlett's tests at the significance level of 0.050. Furthermore, the applicability of the fitted model was verified in each case with diagnostic plots (residual variances, QQ plot, Cook's distance plot). Tukey's honest significant difference (HSD) tests were used as post hoc analyses to assess the significant differences between groups. The data were evaluated using the Kruskal–Wallis test, if the conditions for applying the chosen model were not met, with the use of the Student–Newman–Keuls (SNK) test for the comparison of the different groups at the significance level of 0.050. In addition, the observed hormetic effects were verified with the use of Brain–Cousens hormesis models available in the 'dcr' package of the R Statistical program 4.2.1. [110–112].

3. Results

3.1. Individual and Combined Effects on Algal Growth

During the ecotoxicological testing, significant differences were not observed in the 72 h EC_{50} values determined for the active ingredient glyphosate based on the optical density of *P. subcapitata* (125.2 ± 16.5 mg/L) and *D. subspicatus* (132.9 ± 2.3 mg/L) green algae samples ($p = 0.467$). On the other hand, much higher individual toxicity of glyphosate was demonstrated on *S. obtusiusculus* (73.1 ± 21.2 mg/L) compared to the individual toxicity values determined for the two other green algae species ($p < 0.001$). The individual toxicity of POEA significantly exceeded the individual toxicity of glyphosate for all three algal species ($p < 0.001$). Furthermore, compared to the individual toxicity of glyphosate, the toxicity of the formulation significantly increased in the presence of POEA during the examination of all species (*P. subcapitata, D. subspicatus*: ($p < 0.001$); *S. obtusiusculus* $p = 0.005$). In the case of the formulation, no difference can be observed between the toxicity values corrected with the nominal content of POEA and determined after the individual exposure to POEA for *P. subcapitata* ($p = 0.146$) and *D. subspicatus* ($p = 0.172$), but the individual toxicity of POEA was lower on *S. obtusiusculus* ($p = 0.021$) (Table 4). Based on the determined 72 h EC_{50} values, a significant difference can be observed in the sensitivity of the tested species. There is no difference between the sensitivity of *P. subcapitata* and *D. subspicatus* ($p = 0.467$) for glyphosate, while the sensitivity of *S. obtusiusculus* was higher due to the toxic effects of glyphosate ($p < 0.001$). In contrast, *S. obtusiusculus* was the most tolerant against the effects of the POEA-formulated herbicide and POEA, followed by *D. subspicatus*, while in the case of both tested substances, *P. subcapitata* proved to be the most sensitive green algae (Roundup Classic: $p < 0.001$; POEA: $p < 0.030$) (Table 4).

Table 4. The determined 72 h EC_{50} values for Roundup Classic and its components based on optical density measurements during the ecotoxicological testing on various green algae species.

Algae Species	GLY	72 h EC_{50} Values (mg/L) [1] Roundup Classic [2]			POEA
			GLY cont.	POEA cont.	
Pseudokirchneriella subcapitata	125.2 ± 16.5	12.2 ± 3.1	5.1 ± 1.3	1.9 ± 0.5	2.6 ± 0.7
Desmodesmus subspicatus	132.9 ± 2.3	34.0 ± 6.9	14.1 ± 2.9	5.3 ± 1.1	4.4 ± 0.4
Scenedesmus obtusiusculus	73.1 ± 21.2	65.8 ± 9.0	27.3 ± 3.7	10.2 ± 1.4	6.9 ± 1.6

[1] The combined toxicity of the investigated active ingredient glyphosate (GLY) and formulating agent POEA (mixture of polyethoxylated tallow amines) was investigated in the form of the formulated herbicide preparation. [2] The 72 h EC_{50} values for the herbicide formulation corrected with the nominal content of GLY and POEA indicates the concentration of the given component that is present in the formulation causing a 50% effect.

Similar to the toxicity values based on the optical density measurements, the toxicity of the POEA-formulated herbicide was also higher compared to the individual toxicity of glyphosate on the investigated green algae species ($p < 0.001$), according to the 72 h EC_{50} values based on the measurements of the chlorophyll-a content. The highest individual toxicity of glyphosate (17.4 ± 6.0 mg/L) was demonstrated for the tested cyanobacterium (*A. flos-aquae*) ($p < 0.001$). However, based on the chlorophyll-a content, a difference can be observed in the sensitivity of the two green algae, as *D. subspicatus* proved to be more sensitive (73.8 ± 5.3 mg/L) compared to *P. subcapitata* (105.3 ± 17.8 mg/L) ($p = 0.004$). The individual toxicity of POEA also proved to be much higher compared to the individual toxicity of glyphosate based on the chlorophyll-a content ($p < 0.001$), where *P. subcapitata* was more sensitive to the effect of POEA ($p < 0.001$). In contrast to the active ingredient, there was no significant difference between the 72 h EC_{50} values for the formulation corrected with the POEA content and the determined toxicity values for POEA alone on the tested green algae species ($p \geq 0.096$). Moreover, differences were not indicated in the sensitivity of the green algae species exposed to the formulation ($p = 0.838$) (Table 5).

According to the results of algal growth inhibition tests performed on the APG-formulated herbicide, differences were not demonstrated between the individual toxicity of glyphosate and the combined toxicity indicated by the calculated 72 h EC_{50} values for the formulation on *P. subcapitata* ($p = 0.856$). In contrast to the POEA-formulated GBH, the individual toxicity of glyphosate significantly exceeded the combined toxicity of the components determined in the form of the APG-formulated herbicide on the other two tested green algae species (*D. subspicatus*: $p < 0.001$, *S. obtusiusculus*: $p = 0.002$). *S. obtusiusculus* proved to be more sensitive to the effects of glyphosate ($p = 0.001$) (Table 6). Similar to POEA, the individual toxicity of APG was higher compared to the individual toxicity of glyphosate on *P. subcapitata and D. subspicatus* ($p < 0.001$), where *P. subcapitata* proved to be more sensitive ($p < 0.001$). Conversely, the individual toxicity of glyphosate was higher compared to the individual effects of APG on *S. obtusiusculus* ($p = 0.008$). Significant differences were not detected between the toxicity values determined for Medallon Premium corrected with the APG content and indicated after the individual exposure to the surfactant APG on *P. subcapitata* ($p = 0.068$) and *S. obtusiusculus* ($p = 0.109$), similarly to POEA. However, the individual toxicity of APG was higher compared to the combined effects of the components indicated by toxicity values for the APG-formulated herbicide corrected with the APG content on *D. subspicatus* ($p = 0.001$). Based on the determined 72 h EC_{50} values, significant differences can be observed in the sensitivity of the tested species. *P. subcapitata* was the most sensitive species for both the formulation and APG (Medallon Premium: $p < 0.001$, APG: $p \leq 0.001$). In the case of the additional two green algae, the difference was not demonstrated in their sensitivity to the effects of the APG-formulated herbicide ($p = 0.650$), while *S. obtusiusculus* proved to be more tolerant to the toxic effects of APG ($p < 0.001$) (Table 6).

Table 5. The determined 72 h EC_{50} values for Roundup Classic and its components based on the chlorophyll-a content during the ecotoxicological testing on green algae species and a cyanobacterium.

Algae Species	GLY	72 h EC_{50} Values (mg/L) [1] Roundup Classic [2]			POEA
			GLY cont.	POEA cont.	
Pseudokirchneriella subcapitata	105.3 ± 17.8	34.9 ± 3.2	14.5 ± 1.36	5.4 ± 0.5	1.9 ± 0.3
Desmodesmus subspicatus	73.8 ± 5.3	32.3 ± 9.2	13.4 ± 3.8	5.0 ± 1.4	4.9 ± 0.6
Scenedesmus obtusiusculus	51.1 ± 2.6	25.4 ± 8.5	10.5 ± 3.5	3.9 ± 0.6	4.4 ± 0.9
Anabaena flos-aquae	17.4 ± 6.0	n.m. [3]	n.m.	n.m.	n.m.

[1] The combined toxicity of the investigated active ingredient glyphosate (GLY) and formulating agent POEA (mixture of polyethoxylated tallow amines) was investigated in the form of the formulated herbicide preparation. [2] The 72 h EC_{50} values for the herbicide formulation corrected with the nominal content of GLY and POEA indicates the concentration of the given component that is present in the formulation causing a 50% effect. [3] not measured.

Table 6. The determined 72 h EC_{50} values for Medallon Premium and its components based on the optical density measurements during the ecotoxicological testing on various green algae species.

Algae Species	GLY	72 h EC_{50} Values (mg/L) [1] Medallon Premium [2]			APG
			GLY cont.	APG cont.	
Pseudokirchneriella subcapitata	125.2 ± 16.5	125.7 ± 13.7	42.7 ± 4.7	18.9 ± 2.1	23.0 ± 2.3
Desmodesmus subspicatus	132.9 ± 2.3	720.9 ± 96.6	245.1 ± 32.8	108.1 ± 14.5	64.3 ± 12.9
Scenedesmus obtusiusculus	73.1 ± 21.2	687.5 ± 171.9	233.8 ± 58.4	103.1 ± 25.8	137.9 ± 19.1

[1] The combined toxicity of the investigated active ingredient glyphosate (GLY) and formulating agent APG (alkyl polyglucosides) was investigated in the form of the formulated herbicide preparation. [2] The 72 h EC_{50} values for the herbicide formulation corrected with the nominal content of GLY and APG indicates the concentration of the given component that is present in the formulation causing a 50% effect.

3.2. *Effects on the Photosynthetic Activity of Green Algae Cells*

The individual and combined effects of the components presented in the tested POEA-formulated herbicide on the photosynthetic activity of *P. subcapitata* were evaluated according to the measured Fv*/Fp values connected to the photochemical efficiency of the PS II photochemical system and Rfd* values characterizing photosynthetic activity. During the investigation of the effects of glyphosate on photosynthetic activity, the pure active ingredient did not result in a significant decrease in the Fv*/Fp values compared to the control group up to a concentration of 109 mg/L ($p = 0.034$). In contrast to the pure active ingredient, the formulation resulted in a significant reduction in the Fv*/Fp value ($p = 0.015$) compared to the control group, but only at the highest tested concentration (50.6 mg/L). After the individual exposure to POEA, significant changes in Fv*/Fp values were not observed in the tested concentration range (0.6–19.2 mg/L). POEA in the presence of the active ingredient did not cause the reduction in the Fv*/Fp value up to the highest tested concentration (18.9 mg/L) compared to the control group ($p < 0.001$) (Figure 1—the Fv*/Fp values were plotted in the common concentration range of the tested components: glyphosate: 0–54.5 mg/L; POEA: 0–19.2 mg/L).

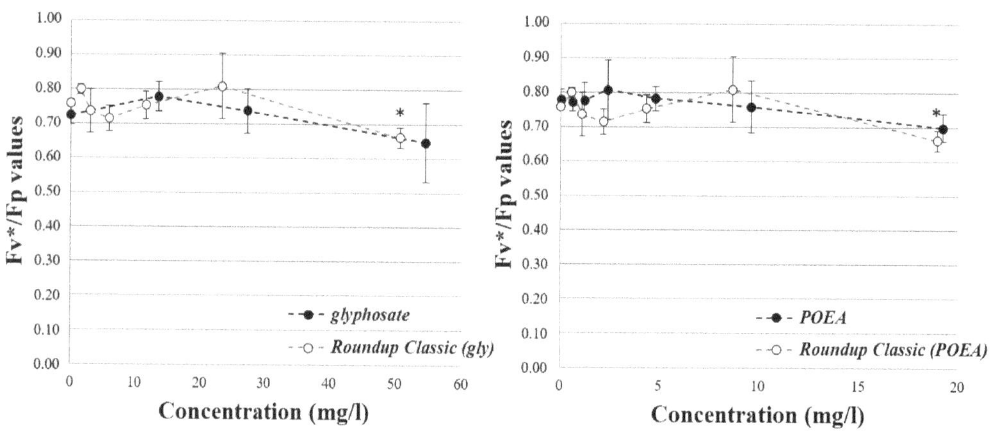

Figure 1. The individual and combined effects of glyphosate and POEA on the Fv*/Fp values characterizing the photochemical efficiency of the PS II photochemical system in the exposed algae cells.

During the investigation of glyphosate, a significant increase was observed in the Rfd* values at the lower tested concentrations (13.6–27.2 mg/L) ($p < 0.025$). However, above this range, no significant difference was observed compared to the control group, not even at the highest concentration (436 mg/L) ($p = 1.000$). After the exposure to glyphosate in the form of herbicide formulation, increased Rfd* values were also detected at the lower test concentrations (1.5–5.8 mg/L) compared to the control group. However, the difference was significant only at the two lowest concentrations ($p < 0.018$). In contrast, significantly decreased Rfd* values were observed at the higher concentration range (11.6–50.6 mg/L) ($p < 0.012$) (Figure 2—the Rfd* values were plotted in the common concentration range of the tested components: glyphosate: 0–54.5 mg/L; POEA: 0–19.2 mg/L). Similar to the effects of glyphosate, a significant increase in Rfd* was observed after the individual exposure to POEA at the lower concentration range (0.6–4.8 mg/L) compared to the control group ($p < 0.035$). However, a significant decrease was demonstrated in the Rfd* values ($p = 0.009$) at the highest tested concentration (19.2 mg/L). After the exposure to the herbicide formulation, POEA also resulted in an increase in Rfd* values at the lower concentration range of POEA (0.5–2.2 mg/L) compared to the control. However, significant differences were not observed at the higher concentrations (4.8–19.2 mg/L) ($p > 0.984$) (Figure 2).

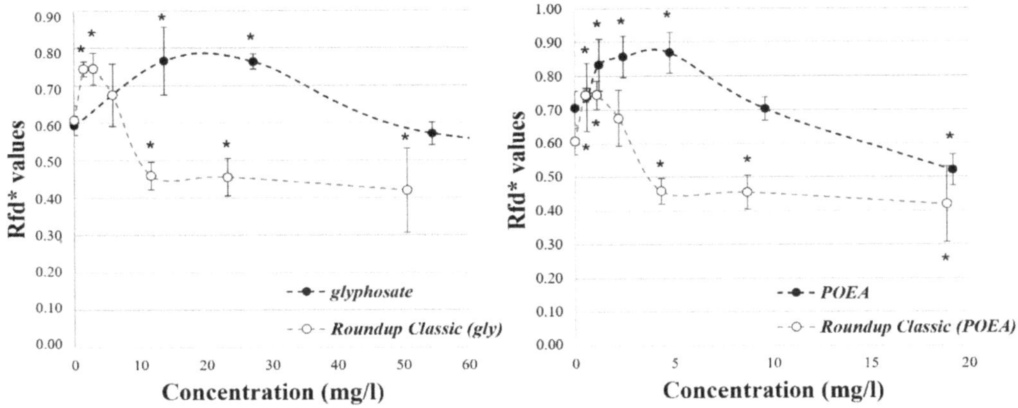

Figure 2. The individual and combined effects of glyphosate and POEA on the Rfd* values characterizing photosynthetic activity in the exposed algae cells.

4. Discussion

According to the determined 72 h EC_{50} values based on the measured optical density and chlorophyll-a content of the samples, the results and observed trends in toxicity correlated well between the tested endpoints. However, higher differences can be observed in some cases. Generally, lower toxicity values were determined based on the chlorophyll-a content compared to the 72 h EC_{50} values calculated based on the optical density of the samples (Tables 4 and 5). The observed differences between the 72 h EC_{50} values based on the two tested endpoints can presumably be explained by the fact that the determination of the optical density can be disturbed by the aggregation of cells and the presence of the remains of dead cells in the sample. On the other hand, during the analytical determination of the chlorophyll-a content, this disturbing matrix is not presented after the extraction of the samples. The determination of chlorophyll-a content proved to be a more reliable and sensitive endpoint, while in dead plant cells, chlorophyll-a begins to decompose rapidly, so the effects of inhibiting algae growth are estimated only based on living cells.

Based on the scientific literature and our results, significant differences can be observed in the sensitivity of different algal and cyanobacterial species to the effects of glyphosate and its formulated herbicides, even within the same taxa [20,70–72,113]. Therefore, the significant differences that can be observed in the available toxicity data are not surprising. Differences in the sensitivity of different algal species can presumably be explained by differences in the morphology of different algal cells (e.g., size and shape of cells, surface area to volume ratio, colony formation), the biology of cells (e.g., cell wall permeability, intracellular structure), and the physiology of different species (e.g., growth, nutrient uptake, metabolic activity) [114,115].

During the examination of the phytotoxic effects of glyphosate, the 72 h EC_{50} values determined for *P. subcapitata* (125.2 $\pm$ 16.5 and 105.3 $\pm$ 17.8 mg/L) far exceed the available literature values (24.7–41 mg/L) [15,73]. The values determined for *D. subspicatus* (132.9 $\pm$ 2.3 and 73.8 $\pm$ 5.3 mg/L) fit well into the available toxicity range (72.9–166 mg/L) [74–76]. The toxicity of GBHs on algae species was investigated in several studies [115–117]. The toxicity values determined for Roundup Classic (corrected with glyphosate content) on *P. subcapitata* (72 h EC_{50} values: 5.1 $\pm$ 1.3 mg/L) correlated well with some of the published 72 h EC_{50} values (0.7–5.8 mg/L) for Roundup formulations [15,73,118]; however, they remain well below the values published in other studies (15.6–64.7 mg/L) [52,119]. In contrast to the 72 h EC_{50} values demonstrated on the MSDS of the tested formulations for algal test organisms (72 h EC_{50} values = 2.1 mg/L (*P. subcapitata*, Roundup Classic) and 140 mg/L (*D. subspicatus*, Medallon Premium) [97,98]), our toxicity values were significantly higher. During the investigation of the individual and combined toxicity of the components presented in Roundup Classic, the formulating agent POEA proved to be the most toxic component, followed by the formulation, while the toxicity of glyphosate was the lowest on the tested green algae species similar to the results of Tsui and Chu [15]. Similarly, the highest toxicity was observed for the tested surfactant APG compared to the individual toxicity of glyphosate and the combined toxic effects of the formulation on *P. subcapitata* and *D. subspicatus*. However, the highest toxicity was observed after the individual glyphosate exposure on *S. obtusiusculus*. (Tables 4–6). The increased toxicity of the formulations in the presence of formulating agents (e.g., POEA) has already been proven in several studies [15,77,120,121]. Based on our results, POEA proved to be more toxic than APG ($p < 0.001$) (Tables 4 and 6). The determined 72 h EC_{50} values of POEA (2.6 ± 0.7 mg/L and 1.9 ± 0.3 mg/L) roughly correspond to the literature data for *P. subcapitata* (0.2–4.1 mg/L) [15,81,82] (Tables 4 and 5). The 72 h EC_{50} values determined for APG on *P. subcapitata* (23.0 ± 2.3 mg/L) correspond to the toxicity range determined for long-chain APG compounds ($C_{12–14}$: 11–46 mg/L) [83,84], but far exceed the value (2.7 mg/L) determined by Pavlic et al. [85] for C_{10} carbon chain APG compounds. In contrast, the 72 h EC_{50} values determined for APG compounds with shorter carbon chains ($C_{8–10}$) on *P. subcapitata* proved to be very high (1113–1543 mg/L) [83,84]. The toxicity values determined for *D. subspicatus* (64.3 ± 12.9 mg/L) (Table 6) far exceed the average 72 h EC_{50} values calculated for APG compounds with different chain lengths ($C_{8–10}$: 21 mg/L, $C_{12–14}$: 6 mg/L) [122], but mainly reflect the average value determined by Pavlic et al. [85] (C_{10}: 0.32 mg/L). According to the results of several studies, increased toxicity on algae species has been observed with the length of the carbon chain [84,122].

In addition to herbicides that directly inhibit photosynthesis (e.g., atrazine), other active ingredients (e.g., glyphosate) can also affect photosynthesis and respiration processes through their effects on different metabolic pathways [90,91,123]. Based on our results, the pure glyphosate active ingredient did not result in a significant decrease in the photochemical efficiency of the PS II photochemical system compared to the control group up to a concentration of 109 mg/L. On the other hand, in the presence of POEA, a significant decrease was detected in the measured Fv*/Fp values even at a lower concentration of 50.6 mg/L (Figure 1). According to the measured Rfd* values, an increase was observed for both the pure active ingredient and the formulated herbicide preparation at low concentrations (glyphosate: 13.6–27.2 mg/L, Roundup Classic (glyphosate equivalent): 1.5–5.8 mg/L). However, the decrease in the Rfd* values was only observed after the exposure to the herbicide formulation in the tested concentration range (Figure 2).

The adverse effects of glyphosate on the photochemical efficiency of the PS II photochemical system were observed on various green algal species (including *P. subcapitata*) from a concentration of 75 mg/L and several diatom species in the range of 15.3–37.5 mg/L [95]. Moreover, similar to our results, at a lower test concentration (0.02 mg/L), an increase in photosynthetic activity was also observed in the unicellular green algae (*Scenedesmus quadricauda*) [113]. The effects of formulating agents (e.g., POEA, APG) on the photosynthetic activity were investigated on the leaves of higher order plants (*Brassica oleracea*, *Malus domestica*), where significant effects of the tested surfactants were not demonstrated on *M. domestica*. On the other hand, a significant decrease in the photochemical efficiency of the PS II photochemical system was detected on the leaves of *B. oleracea* exposed to POEA [124]. During our measurements, POEA caused a significant decrease in the photochemical efficiency of the PS II photochemical system only in the presence of glyphosate at the highest test concentration (Roundup Classic: 18.9 mg/L POEA equivalent). POEA individually and in the presence of glyphosate induced an increase in Rfd* values at lower test concentrations. Conversely, significantly lower Rfd* values were demonstrated in the higher POEA concentration range compared to the control (Figure 2). After the exposure to Roundup formulations, the phytotoxic effect of glyphosate in the presence of POEA far exceeded the toxicity of the pure active ingredient on the photosynthetic activity of green and blue-green algae species (*M. aeruginosa*, *N. microcarpa* *var. wrightii*). However, at low concentrations, an increase in photosynthetic activity was also reported [80,90].

The observed stimulatory effects of toxic compounds at lower concentrations can be interpreted as hormesis effects [80,90]. Hormesis is the phenomenon when a xenobiotic has an opposite effect in low and high doses on some infra-individual level property of an organism (biochemical processes, cellular characteristics, histological and organic changes, etc.), or on some characteristic of a population/community [125]. Essentially, hormesis is a biological phenomenon where a harmful compound shows a favorable or stimulating effect in the low concentration range [126]. The phenomenon cannot be characterized by the usual sigmoid (logistic) shaped dose–effect curve [126–128]. The explanation of the phenomenon of hormesis is not completely clear, but several background mechanisms can be assumed [127,129]. One of the possible explanations is that the homeostasis of the organism is disturbed by the low concentration of the pollutant and the positive effect appears to compensate the negative effects of the xenobiotics. As a result of the disturbance, the dynamic equilibrium of the body conditions slightly exceeds the normal limits. To compensate for this imbalance, the affected organism mobilizes resources and, in the meantime, achieves a more favorable state than before (e.g., observed higher growth rates by the low concentrations of the tested compounds compared to the control) [127,130]. According to another idea, hormesis results from changes in energy allocation of the organisms [131]. The consequence of the trade-off alterations in life history traits can be indicated by changes in various population parameters (e.g., number of eggs, growth, and behavior) [129,131]. The changes of trade-off caused by pesticides can have an adaptive value, because it helps individuals maintain their fitness [132].

Based on our results, Rfd* proved to be a more sensitive endpoint compared to the Fv*/Fp values characterizing the photochemical efficiency of the PS II photochemical system. During the investigation of the phytotoxic effects, a higher growth rate (3.5 mg/L Roundup Classic) and increased photosynthetic activity were measured at lower test concentrations of the formulation compared to the control groups. At the lower concentrations of the pure active ingredient and POEA, the hormesis effects were only detected during the measurement of the Rfd* values. The observed hormetic effects were also verified by the performed Brain–Cousens hormesis models ($p < 0.048$). Based on the results of algal growth inhibition tests, this stimulating effect was not demonstrated after the individual exposure to the tested components. According to the measured Rfd* values, the hormetic response of *P. subcapitata* was indicated after the individual and combined exposure to the components at the lower concentration ranges. In our study, the observed changes in photosynthetic parameters can presumably be explained primarily by the phenomenon of hormesis, as well as by the change in algal biomass resulting from toxic effects. At low concentrations, the toxic effects do not yet prevail, but on the contrary, the treated algal cells can utilize glyphosate as a source of carbon, phosphorus, and nitrogen [70,90,110,133,134]. Moreover, glyphosate can also trigger pathways for protein and metabolite synthesis [70,133], which can result in increased biomass growth. However, with the increase in concentration, the toxic effects prevail against the excess nutrient content. During the investigation of *Pseudomonas* species, the utilization of octadecyl-bis(2-hydroxyethyl) amine was demonstrated as a carbon and energy source during bacterial growth [81]. Based on certain studies, hormesis can also be interpreted as the response of the plant organism to increased stress [135].

Significant differences were demonstrated in the individual and combined toxicity of the components presented in the tested GBHs. Furthermore, our results support the scientific opinions proposing changes in the official regulations, including the strict regulation of co-formulants, the future development of standards to assess combined effects, and the environmental risks of chemical mixtures [136,137]. Generally, the active ingredients and the co-formulants almost certainly become separated relatively quickly after pesticide treatments. The mobility of the components presented in pesticide formulation highly depends on the physico-chemical properties of the chemical substances (e.g., water solubility, log K_{OW}) and the environmental matrices (e.g., pH, level of suspended materials, dissolved oxygen content) [20–22,41,138]. The water solubility of glyphosate is 11.6 g/L (25 °C), while degradation half-life (DT_{50}) in water varies from a few to 91 days [41]. The water solubility of POEA increases with the increase in oxide–tallow amine ratio [139], while its persistence was demonstrated in soil by several studies [44,45]. The water solubility and biodegradability of APGs depends on the length of the alkyl chain [48]. Various co-formulants affect the solubility and stability of glyphosate, leading to variations in its bioavailability and persistence in the environment [43]. The surfactants applied in GBHs can modify the adsorption capacity of glyphosate, resulting in reduced physical adsorption of glyphosate on the surface of solid–liquid boundary phases (e.g., suspended particles in water samples) [140]. In addition, surfactants form micelles that help glyphosate stay in solution and provide protection against degradation [82,141,142]. In summary, the physicochemical interactions between glyphosate and the additional ingredients are complex and can significantly influence the overall toxicity of GBHs. Therefore, the ecotoxicological and toxicological evaluation of the various additives is an essential condition for the proper environmental risk assessment of pesticide formulations used in agricultural practice. Currently, manufacturers are only required to indicate the exact chemical name and quantity of the active ingredient(s), synergists, and antidotes on the labels of the products in the EU. Thus, the exact composition of the formulations and information about co-formulants are not public [9,143], resulting in several uncertainties regarding the evaluation of the possible combined toxic effects [144].

However, most of the calculated toxicity values determined for the components presented in glyphosate-based formulations individually and in combination remain below the detected average environmental concentrations in surface waters [27,36], and contamination levels can rise significantly after heavy rains in the watercourses near the treated areas [34,145]. Additionally, the toxicity of glyphosate and POEA can also be significantly influenced by different environmental conditions (e.g., pH, dissolved oxygen content, temperature) [146,147]. Moreover, as a result of global climate change, increased average temperature and the modified intensity of the incident light can significantly influence the phytotoxic effects of glyphosate on phytoplankton communities [91]. Stimulatory effects were indicated at lower concentrations of glyphosate-based Roundup Classic (1.5–5.8 mg glyphosate/L and 0.5–2.2 mg POEA/L), for which concentrations approach or stay much below the measured maximums [27,34,35,37].

5. Conclusions

Based on the scientific evidence and our findings, significant differences can be observed in both the individual and combined toxicity of the components contained in the tested GBH formulations. According to our results, the tested co-formulants proved to be the most toxic components. Although the individual toxicity of APG is not as high as for POEA, the toxicity of the formulation is affected by the simultaneous presence of the active ingredient and the co-formulants. Therefore, the revision of GBHs formulated with APG compounds may also be necessary. In addition, significant differences were detected in the sensitivity of the tested algal species, including *D subspicatus* and *S. obtusiusculus* species belonging to the same family (*Scenedesmaceae*). The differences in sensitivity are presumably the result of differences observed in the morphology, cell biology, and physiology of different algal cells. During the evaluation of phytotoxic effects, increased photosynthetic activity was detected on *P. subcapitata* after the exposure to the POEA-formulated GBH and its components in the lower concentration ranges. However, decreased activity was observed after exposure to POEA and the formulation at higher test concentrations. Not only the inhibitory effects but stimulating effects on the growth of algae can adversely affect the aquatic ecosystem and water quality of surface waters. Moreover, the accumulation of phytotoxins can also cause serious environmental effects on aquatic communities.

Author Contributions: Conceptualization, S.K.; methodology, S.K., E.T., A.B. and S.L.; software, A.B. and S.L.; validation, S.K. and E.T.; investigation, S.K., E.T. and B.D.; writing—original draft preparation, S.K. and E.T.; writing—review and editing, A.B., B.D., L.K. and A.S.; visualization,

S.K.; supervision, A.S.; funding acquisition, L.K. and A.S. All authors have read and agreed to the published version of the manuscript.

Funding: This work has been funded by the Hungarian National Research, Development and Innovation Office within the National Competitiveness and Excellence Program, project NVKP_16-1-2016-0049, "In situ, complex water quality monitoring by using direct or immunofluorimetry and plasma spectroscopy" (Aquafluosense 2017–2021), and project 2022-2.1.1-NL-2022-00006, "Development of the Agrotechnology National Laboratory" (Grant agreement NKFIH-3524-1/2022), supported by the National Research, Development and Innovation Fund by the Hungarian Ministry of Culture and Innovation, as well as projects TKP2021-NVA-02 and TKP2021-NVA-22 within the framework of the Thematic Excellence Program 2021, National Defense, National Security Sub-Program. Our study was also supported by the Hungarian Scientific Research Fund (OTKA K109865 and OTKA K112978 projects), the Research Excellence Program 2024 of Hungarian University of Agriculture and Life Sciences, the János Bolyai Scholarship of the Hungarian Academy of Sciences, and ÚNKP-23-5-BME-448 New National Excellence Program of the Ministry for Culture and Innovation from the source of the National Research, Development and Innovation Fund.

Institutional Review Board Statement: Not applicable.

Informed Consent Statement: Not applicable.

Data Availability Statement: The data presented in this study are available on request from the corresponding author. The data are not publicly available due to privacy reasons.

Acknowledgments: The authors thank Péter Bohus for the sample of testing materials and express their sincere appreciation to Ildikó Mándics and Judit Olajos for their technical assistance during ecotoxicological testing.

Conflicts of Interest: The authors declare no conflicts of interest.

References

1. Arias-Estévez, M.; López-Periago, E.; Martínez-Carballo, E.; Simal-Gándara, J.; Mejuto, J.-C.; García-Río, L. The mobility and degradation of pesticides in soils and the pollution of groundwater resources. *Agric. Ecosyst. Environ.* **2008**, *123*, 247–260. [CrossRef]
2. Stehle, S.; Schulz, R. Agricultural insecticides threaten surface waters at the global scale. *Proc. Natl. Acad. Sci. USA* **2015**, *112*, 5750–5755. [CrossRef]
3. Tang, F.H.M.; Lenzen, M.; McBratney, A.; Maggi, F. Risk of pesticide pollution at the global scale. *Nat. Geosci.* **2021**, *14*, 206–210. [CrossRef]
4. Solomon, K.R.; Thompson, D.G. Ecological risk assessment for aquatic organisms from over-water uses of glyphosate. *J. Toxicol. Environ. Health Part B* **2003**, *6*, 289–324. [CrossRef] [PubMed]
5. Hanke, I.; Wittmer, I.; Bischofberger, S.; Stamm, C.; Singer, H. Relevance of urban glyphosate use for surface water quality. *Chemosphere* **2010**, *81*, 422–429. [CrossRef] [PubMed]
6. Benbrook, C.M. Trends in glyphosate herbicide use in the United States and globally. *Environ. Sci. Eur.* **2016**, *28*, 3. [CrossRef]
7. Gower, S.A.; Loux, M.M.; Cardina, J.; Harrison, S.K. Effect of planting date, residual herbicide, and postemergence application timing on weed control and grain yield in glyphosate-tolerant corn (*Zea mays*). *Weed Technol.* **2002**, *16*, 488–494. [CrossRef]
8. Whigham, D.K.; Stoller, E.W. Soybean desiccation by paraquat, glyphosate, and ametryn to accelerate harvest. *Agron. J.* **1979**, *71*, 630–633. [CrossRef]
9. European Parliament and Council. Regulation (EC) No 1107/2009 of the European Parliament and of the Council of 21 October 2009 concerning the placing of plant protection products on the market and repealing council directives 79/117/EEC and 91/414/EEC. *OJ EU* **2009**, *L309*, 1–50. Available online: https://eur-lex.europa.eu/eli/reg/2009/1107/oj (accessed on 21 March 2024).
10. European Commission. Commission Implementing Regulation (EU) 2023/2660 of 28 November 2023 renewing the approval of the active substance glyphosate in accordance with Regulation (EC) No 1107/2009 of the European Parliament and of the Council and amending Commission Implementing Regulation (EU) No 540/2011. OJ EU 2023, L2023/2660. Available online: https://eur-lex.europa.eu/legal-content/EN/TXT/HTML/?uri=OJ:L_202302660&qid=1710855398897 (accessed on 21 March 2024).
11. Defarge, N.; Takács, E.; Lozano, V.L.; Mesnage, R.; Vendômois, J.S.; Seralini, G.E.; Székács, A. Co-formulants in glyphosate-based herbicides disrupt aromatase activity in human cells below toxic levels. *Int. J. Environ. Res. Pub. Health* **2016**, *13*, 264. [CrossRef]
12. Travlos, I.; Cheimona, N.; Bilalis, D. Glyphosate efficacy of different salt formulations and adjuvant additives on various weeds. *Agronomy* **2017**, *7*, 60. [CrossRef]
13. Foy, C. Adjuvants: Terminology, classification, and mode of action. In *Adjuvants and Agrochemicals*; Chow, P., Grant, C., Hinshalwood, A., Simundson, E., Eds.; CRC Press: Boca Raton, FL, USA, 1987; pp. 1–15.
14. Defarge, N.; Spiroux de Vendômois, J.; Séralini, G.E. Toxicity of formulants and heavy metals in glyphosate-based herbicides and other pesticides. *Toxicol. Rep.* **2018**, *5*, 156–163. [CrossRef] [PubMed]

15. Tsui, M.T.K.; Chu, L.M. Aquatic toxicity of glyphosate-based formulations: Comparison between different organisms and the effects of environmental factors. *Chemosphere* **2003**, *52*, 1189–1197. [CrossRef] [PubMed]
16. Székács, A. Mechanism-related teratogenic, hormone modulant and other toxicological effects of veterinary and agricultural surfactants. *Insights Vet. Sci.* **2017**, *1*, 24–31. [CrossRef]
17. Mesnage, R.; Benbrook, C.; Antoniou, M.N. Insight into the confusion over surfactant co-formulants in glyphosate-based herbicides. *Food Chem. Toxicol.* **2019**, *128*, 137–145. [CrossRef] [PubMed]
18. European Commission. Commission Implementing Regulation (EU) 2016/1313 of 1 August 2016 amending Implementation Regulation (EU) No 540/2011 as regards the conditions of approval of the active substance glyphosate. *OJ EU* **2016**, *L208*, 1–3. Available online: https://eur-lex.europa.eu/legal-content/EN/TXT/?uri=uriserv:OJ.L_.2016.208.01.0001.01.ENG (accessed on 21 March 2024).
19. Székács, A.; Mörtl, M.; Darvas, B. Monitoring pesticide residues in surface and ground water in Hungary: Surveys in 1990–2015. *J. Chem.* **2015**, *2015*, 717948. [CrossRef]
20. Klátyik, S.; Simon, G.; Oláh, M.; Takács, E.; Mesnage, R.; Antoniou, M.N.; Zaller, J.G.; Székács, A. Aquatic ecotoxicity of glyphosate, its formulations, and co-formulants: Evidence from 2010 to 2023. *Environ. Sci. Eur.* **2024**, *36*, 22. [CrossRef]
21. Borggaard, O.K.; Gimsing, A.L. Fate of glyphosate in soil and the possibility of leaching to ground and surface waters: A review. *Pest Manag. Sci.* **2008**, *64*, 441–456. [CrossRef] [PubMed]
22. Hébert, M.-P.; Fugère, V.; Gonzalez, A. The overlooked impact of rising glyphosate use on phosphorus loading in agricultural watersheds. *Front. Ecol. Environ.* **2019**, *17*, 48–56. [CrossRef]
23. Zaller, J.G.; Weber, M.; Maderthaner, M.; Gruber, E.; Takács, E.; Mörtl, M.; Klátyik, S.; Győri, J.; Römbke, J.; Leisch, F.; et al. Effects of glyphosate-based herbicides and their active ingredients on earthworms, water infiltration and glyphosate leaching are influenced by soil properties. *Environ. Sci. Eur.* **2021**, *33*, 51. [CrossRef]
24. Kjær, J.; Olsen, P.; Ullum, M.; Grant, R. Leaching of glyphosate and amino-methylphosphonic acid from Danish agricultural field sites. *J. Environ. Qual.* **2005**, *34*, 608–620. [CrossRef] [PubMed]
25. Huhn, C. More and enhanced glyphosate analysis is needed. *Anal. Bioanal. Chem.* **2018**, *410*, 3041–3045. [CrossRef] [PubMed]
26. Duke, S.O.; Powles, S.B. Glyphosate: A once-in-a-century herbicide. *Pest. Manag. Sci.* **2008**, *64*, 319–325. [CrossRef] [PubMed]
27. Székács, A.; Darvas, B. Re-registration challenges of glyphosate in the European Union. *Front. Environ. Sci.* **2018**, *6*, 78. [CrossRef]
28. Villeneuve, A.; Larroudé, S.; Humbert, J.F. Herbicide contamination of freshwater ecosystems: Impact on microbial communities. In *Pesticides–Formulations, Effects, Fate*; Stoytcheva, M., Ed.; InTech: Rijeka, Croatia, 2011; pp. 285–312.
29. Chang, F.C.; Simcik, M.F.; Capel, P.D. Occurrence and fate of the herbicide glyphosate and its degradate aminomethylphosphonic acid in the atmosphere. *Environ. Toxicol. Chem.* **2011**, *30*, 548–555. [CrossRef] [PubMed]
30. Silva, V.; Montanarella, L.; Jones, A.; Fernández-Ugalde, O.; Mol, H.G.J.; Ritsema, C.J.; Geissen, V. Distribution of glyphosate and aminomethylphosphonic acid (AMPA) in agricultural topsoils of the European Union. *Sci. Total Environ.* **2018**, *621*, 1352–1359. [CrossRef] [PubMed]
31. Lutri, V.F.; Matteoda, E.; Blarasin, M.; Aparicio, V.; Giacobone, D.; Maldonado, L.; Becher Quinodoz, F.; Cabrera, A.; Giuliano Albo, J. Hydrogeological features affecting spatial distribution of glyphosate and AMPA in groundwater and surface water in an agroecosystem. Córdoba, Argentina. *Sci. Total Environ.* **2020**, *711*, 134557. [CrossRef]
32. Grandcoin, A.; Piel, S.; Baurès, E. AminoMethylPhosphonic acid (AMPA) in natural waters: Its sources, behavior and environmental fate. *Weed Res.* **2017**, *117*, 187–197. [CrossRef]
33. U.S. Geological Survey. Common Weed Killer Is Widespread in the Environment, U.S. Geological Survey. Available online: https://www.usgs.gov/programs/environmental-health-program/science/common-weed-killer-widespread-environment (accessed on 21 March 2024).
34. Coupe, R.H.; Kalkhoff, S.J.; Capel, P.D.; Gregoire, C. Fate and transport of glyphosate and aminomethylphosphonic acid in surface waters of agricultural basins. *Pest Manag. Sci.* **2012**, *68*, 16–30. [CrossRef]
35. Mörtl, M.; Németh, G.; Juracsek, J.; Darvas, B.; Kamp, L.; Rubio, F.; Székács, A. Determination of glyphosate residues in Hungarian water samples by immunoassay. *Microchem. J.* **2013**, *107*, 143–151. [CrossRef]
36. Bonansea, R.I.; Filippi, I.; Wunderlin, D.A.; Marino, D.J.G.; Amé, M.V. The fate of glyphosate and AMPA in a freshwater endorheic basin: An ecotoxicological risk assessment. *Toxics* **2018**, *6*, 3. [CrossRef] [PubMed]
37. Caprile, A.C.; Aparicio, V.; Sasal, C.; Andriulo, E. Variation in glyphosate and AMPA concentrations of surface water and groundwater. *Geophys. Res. Abstr.* **2017**, *19*, EGU2017-2068. Available online: https://meetingorganizer.copernicus.org/EGU2017/EGU2017-2068.pdf (accessed on 21 March 2024).
38. Centrum voor Milieuwetenschappen Leiden. Atlas Bestrijdingsmiddelen in Oppervlaktewater, Universiteit Leiden, Leiden, Netherlands. Available online: https://www.bestrijdingsmiddelenatlas.nl/atlas/1/1 (accessed on 21 March 2024).
39. Poiger, T.; Buerge, I.J.; Bächli, A.; Müller, M.D.; Balmer, M.E. Occurrence of the herbicide glyphosate and its metabolite AMPA in surface waters in Switzerland determined with on-line solid phase extraction LC-MS/MS. *Environ. Sci. Pollut. Res.* **2017**, *24*, 1588–1596. [CrossRef] [PubMed]
40. Di Guardo, A.; Finizio, A. A new methodology to identify surface water bodies at risk by using pesticide monitoring data: The glyphosate case study in Lombardy Region (Italy). *Sci. Total Environ.* **2018**, *610–611*, 421–429. [CrossRef]
41. Turner, J.A. *The Pesticide Manual*, 19th ed.; The British Crop Protection Council: Brighton, UK, 2021.

42. Singh, S.; Kumar, V.; Gill, J.P.K.; Datta, S.; Singh, S.; Dhaka, V.; Kapoor, D.; Wani, A.B.; Dhanjal, D.S.; Kumar, M.; et al. Herbicide glyphosate: Toxicity and microbial degradation. *Int. J. Environ. Res. Public Health.* **2020**, *15*, 7519. [CrossRef]
43. Klátyik, S.; Simon, G.; Oláh, M.; Mesnage, R.; Antoniou, M.N.; Zaller, J.G.; Székács, A. Terrestrial ecotoxicity of glyphosate, its formulations, and co-formulants: Evidence from 2010–2023. *Environ. Sci. Eur.* **2023**, *35*, 51. [CrossRef]
44. Tush, D.; Meyer, M.T. Polyoxyethylene tallow amine, a glyphosate formulation adjuvant: Soil adsorption characteristics, degradation profile, and occurrence on selected soils from agricultural fields in Iowa, Illinois, Indiana, Kansas, Mississippi, and Missouri. *Environ. Sci. Technol.* **2016**, *50*, 5781–5789. [CrossRef] [PubMed]
45. Tush, D.; Maksimowicz, M.M.; Meyer, M.T. Dissipation of polyoxyethylene tallow amine (POEA) and glyphosate in an agricultural field and their co-occurrence on streambed sediments. *Sci. Total Environ.* **2018**, *636*, 212–219. [CrossRef]
46. Morrás, H.; Behrends Kraemer, F.; Sainz, D.; Fernández, P.; Chagas, C. Soil structure and glyphosate fate under no-till management in the Pampa region. II. Glyphosate and AMPA persistence and spatial distribution in the long-term. A conceptual model. *Soil Tillage Res.* **2022**, *223*, 105471. [CrossRef]
47. Green, J.M.; Beestman, G.B. Recently patented and commercialized formulation and adjuvant technology. *Crop Prot.* **2007**, *26*, 320–327. [CrossRef]
48. Geetha, D.; Tyagi, R. Alkyl poly glucosides (APGs) surfactants and their properties: A review. *Tenside Surfactants Deterg.* **2012**, *49*, 417–427. [CrossRef]
49. Rastogi, R. Fate of alkyl polyglucosides in the environment. *J. Cosmet. Sci.* **2021**, *72*, 91–98. [PubMed]
50. Evalen, P.S.; Barnhardt, E.N.; Ryu, J.; Stahlschmidt, Z.R. Toxicity of glyphosate to animals: A meta-analytical approach, *Environ. Pollut.* **2024**, *347*, 123669. [CrossRef] [PubMed]
51. Dabney, B.L.; Patiño, R. Low-dose stimulation of growth of the harmful alga, *Prymnesium parvum*, by glyphosate and glyphosate-based herbicides. *Harmful Algae* **2018**, *80*, 130–139. [CrossRef] [PubMed]
52. Fernández, C.; Asselborn, V.; Parodi, E.R. Toxic effects of chlorpyrifos, cypermethrin and glyphosate on the non-target organism *Selenastrum capricornutum* (*Chlorophyta*). *An. Acad. Bras. Cienc.* **2021**, *93*, e20200233. [CrossRef] [PubMed]
53. Reno, U.; Doyle, S.R.; Momo, F.R.; Regaldo, L.; Gagneten, A.M. Effects of glyphosate formulations on the population dynamics of two freshwater cladoceran species. *Ecotoxicology* **2018**, *27*, 784–793. [CrossRef]
54. Iummato, M.M.; Sabatini, S.E.; Cacciatore, L.C.; Cochón, A.C.; Cataldo, D.; de Molina, M.D.C.R.; Juárez, Á.B. Biochemical responses of the golden mussel *Limnoperna fortunei* under dietary glyphosate exposure. *Ecotoxicol. Environ. Saf.* **2018**, *163*, 69–75. [CrossRef] [PubMed]
55. Fiorino, E.; Sehonova, P.; Plhalova, L.; Blahova, J.; Svobodova, Z.; Faggio, C. Effects of glyphosate on early life stages: Comparison between *Cyprinus carpio* and *Danio rerio*. *Environ. Sci. Pollut. Res.* **2018**, *25*, 8542–8549. [CrossRef]
56. Bach, N.C.; Marino, D.J.G.; Natale, G.S.; Somoza, G.M. Effects of glyphosate and its commercial formulation, Roundup® Ultramax, on liver histology of tadpoles of the neotropical frog, *Leptodactylus latrans* (amphibia: *Anura*). *Chemosphere* **2018**, *202*, 289–297. [CrossRef]
57. Schaffer, J.D.; Sebetich, M.J. Effects of aquatic herbicides on primary productivity of phytoplankton in the laboratory. *Bull. Environ. Contam. Toxicol.* **2004**, *72*, 1032–1037. [CrossRef] [PubMed]
58. Jyothi, K.; Krishna Prasad, M.; Mohan Narasimha Rao, G. Algae in fresh water ecosystem. *Phykos* **2016**, *46*, 25–31.
59. Maguire, R.J.; Wong, P.T.S.; Rhamey, J.S. Accumulation and metabolism of tri-n-butyltin cation by a green alga, *Ankistrodesmus falcatus*. *Can. J. Fish. Aquat.* **1984**, *41*, 537–540. [CrossRef]
60. Paerl, H.W.; Otten, T.G. Harmful cyanobacterial blooms: Causes, consequences, and controls. *Microb. Ecol.* **2013**, *65*, 995–1010. [CrossRef] [PubMed]
61. Watson, S.B.; Whitton, B.A.; Higgins, S.N.; Paerl, H.W.; Brooks, B.W.; Wehr, J.D. Harmful algal blooms. In *Freshwater Algae of North America: Ecology and Classification*; Wehr, J.D., Sheath, R.G., Kociolek, P., Eds.; Academic Press: Cambridge, MA, USA, 2015; pp. 873–920.
62. Wu, N.; Dong, X.; Liu, Y.; Wang, C.; Baattrup-Pederson, A.; Riis, T. Using river microalgae as indicators for freshwater biomonitoring: Review of published research and future directions. *Ecol. Indic.* **2017**, *81*, 124–131. [CrossRef]
63. Vidyashankar, S.; Ravishankar, G.A. Algae-based bioremediation: Bioproducts and biofuels for biobusiness. In *Bioremediation and Bioeconomy*; Prasad, M.N.V., Ed.; Elsevier: Amsterdam, The Netherlands, 2016; pp. 457–493.
64. Qu, M.; Wang, L.; Xu, Q.; An, J.; Mei, Y.; Liu, G. Influence of glyphosate and its metabolite aminomethylphosphonic acid on aquatic plants in different ecological niches. *Ecotoxicol. Environ. Saf.* **2022**, *246*, 114155. [CrossRef] [PubMed]
65. Hernández-García, C.I.; Martínez-Jerónimo, F. Multistressor negative effects on an experimental phytoplankton community. The case of glyphosate and one toxigenic cyanobacterium on *Chlorophycean* microalgae. *Sci. Total Environ.* **2020**, *717*, 137186. [CrossRef] [PubMed]
66. Wu, L.; Qiu, Z.; Zhou, Y.; Du, Y.; Liu, C.; Ye, J.; Hu, X. Physiological effects of the herbicide glyphosate on the cyanobacterium *Microcystis aeruginosa*. *Aquat. Toxicol.* **2016**, *178*, 72–79. [CrossRef] [PubMed]
67. Drzyzga, D.; Lipok, J. Glyphosate dose modulates the uptake of inorganic phosphate by freshwater cyanobacteria. *J. Appl. Phycol.* **2018**, *30*, 299–309. [CrossRef]
68. Iummato, M.M.; Fassiano, A.; Graziano, M.; Dos Santos Afonso, M.; Ríos de Molina, M.D.C.; Juárez, Á.B. Effect of glyphosate on the growth, morphology, ultrastructure and metabolism of *Scenedesmus vacuolatus*. *Ecotoxicol. Environ. Saf.* **2019**, *15*, 471–479. [CrossRef]

69. Smedbol, É.; Gomes, M.P.; Paquet, S.; Labrecque, M.; Lepage, L.; Lucotte, M.; Juneau, P. Effects of low concentrations of glyphosate-based herbicide Factor 540® on an agricultural stream freshwater phytoplankton community. *Chemosphere* **2018**, *192*, 133–141. [CrossRef] [PubMed]
70. Wang, C.; Lin, X.; Li, L.; Lin, S. Differential growth responses of marine phytoplankton to herbicide glyphosate. *PLoS ONE* **2016**, *11*, e0151633. [CrossRef] [PubMed]
71. Ma, J.; Qin, W.; Lu, N.; Wang, P.; Huang, C.; Xu, R. Differential sensitivity of three cyanobacteria (*Anabaena flos-aquae, Microcystis flos-aquae* and *Mirocystis aeruginosa* to 10 pesticide adjuvants. *Bull. Environ. Contam. Toxicol.* **2005**, *75*, 873–881. [CrossRef] [PubMed]
72. Arunakumara, K.; Walpola, B.; Yoon, M. Metabolism and degradation of glyphosate in aquatic cyanobacteria: A review. *Afr. J. Microbiol. Res.* **2013**, *7*, 4084–4090.
73. EGEIS Aquatic Ecotoxicity of Glyphosate and Formulated Products Containing Glyphosate. European Glyphosate Environmental Information Sources. Available online: http://www.egeis.org/cd-info/Aquatic-ecotoxicity-of-glyphosate-and-formulated-products-containing-glyphosate.pdf (accessed on 21 March 2024).
74. Giesy, J.P.; Dobson, S.; Solomon, K.R. Ecotoxicological risk assessment for Roundup herbicide. *Rev. Environ. Contam. Toxicol.* **2000**, *167*, 35–120.
75. Lewis, K.A.; Tzilivakis, J.; Warner, D.J.; Green, A. An international database for pesticide risk assessments and management. *Hum. Ecol. Risk. Assess.* **2016**, *22*, 1050–1064. [CrossRef]
76. MacBean, C. *The Pesticide Manual*, 16th ed.; The British Crop Protection Council: Brighton, UK, 2012.
77. Gonzalez, D.; Juárez, A.; Krug, C.; Santos, M.; Vera, S. Freshwater periphyton response to technical-grade and two commercial formulations of glyphosate. *Ecol. Austral.* **2019**, *29*, 020–027. [CrossRef]
78. Vera, M.S.; Trinelli, M.A. First evaluation of the periphyton recovery after glyphosate exposure. *Environ. Pollut.* **2021**, *290*, 117998. [CrossRef] [PubMed]
79. Bricheux, G.; Le Moal, G.; Hennequin, C.; Coffe, G.; Donnadieu, F.; Portelli, C.; Bohatier, J.; Forestier, C. Characterization and evolution of natural aquatic biofilm communities exposed in vitro to herbicides. *Ecotoxicol. Environ. Saf.* **2013**, *88*, 126–134. [CrossRef]
80. de Campos Oliveira, R.; Boas, L.K.V.; Branco, C.C.Z. Assessment of the potential toxicity of glyphosate-based herbicides on the photosynthesis of *Nitella microcarpa* var. *wrightii (Charophyceae)*. *Phycologia* **2016**, *55*, 577–584. [CrossRef]
81. van Ginkel, C.G.; Stroo, C.A.; Kroon, A.G.M. Biodegradability of ethoxylated fatty amines and amides and the non-toxicity of their biodegradation products. *Tenside Surfactant. Deterg.* **1993**, *30*, 213–216. [CrossRef]
82. Rodriguez-Gil, J.L.; Prosser, R.; Poirier, D.; Lissemore, L.; Thompson, D.; Hanson, M.; Solomon, K.R. Aquatic hazard assessment of MON 0818, a commercial mixture of alkylamine ethoxylates commonly used in glyphosate-containing herbicide formulations. Part 1: Species sensitivity distribution from laboratory acute exposures. *Environ. Toxicol. Chem.* **2017**, *36*, 512–521. [CrossRef] [PubMed]
83. Madsen, T.; Petersen, G.; Seiero, C.; Torslov, J. Biodegradability and aquatic toxicity of glycoside surfactants and a nonionic alcohol ethoxylate. *J. Am. Oil Chem. Soc.* **1996**, *73*, 929–933. [CrossRef]
84. Jurado, E.; Fernández-Serrano, M.; Núnez-Olea, J.; Lechuga, M.; Jiménez, J.L.; Ríos, F. Acute toxicity of alkylpolyglucosides to *Vibrio fischeri, Daphnia magna* and microalgae: A comparative study. *Bull. Environ. Contam. Toxicol.* **2012**, *88*, 290–295. [CrossRef] [PubMed]
85. Pavlic, Z.; Vidakovic-Cifrek, Z.; Puntaric, D. Toxicity of surfactants to green microalgae *Pseudokirchneriella subcapitata* and *Scenedesmus subspicatus* and to marine diatoms *Phaeodactylum tricornutum* and *Skeletonema costatum*. *Chemosphere* **2005**, *61*, 1061–1068. [CrossRef]
86. Kautsky, H.; Hirsch, A. Neue Versuche zur Kohlenstoffassimilation. *Naturwissenschaften* **1931**, *19*, 964. [CrossRef]
87. Krause, G.H.; Weis, E. Chlorophyll fluorescence as a tool in plant physiology II: Interpretation of fluorescence signal. *Photosynth. Res.* **1984**, *5*, 139–157. [CrossRef]
88. Barócsi, A.; Kocsányi, L.; Várkonyi, S.; Richter, P.; Csintalan, Z.; Szente, K. Two-wavelength, multipurpose, truly portable chlorophyll fluorometer and its application in field monitoring of phytoremediation. *Meas. Sci. Technol.* **2000**, *11*, 717–729. [CrossRef]
89. Lenk, S.; Gádoros, P.; Kocsányi, L.; Barócsi, A. Teaching laser-induced fluorescence of plant leaves. *Eur. J. Phys.* **2016**, *37*, 064003. [CrossRef]
90. Qiu, H.; Geng, J.; Ren, H.; Xia, X.; Wang, X.; Yu, Y. Physiological and biochemical responses of *Microcystis aeruginosa* to glyphosate and its Roundup® formulation. *J. Hazard. Mater.* **2013**, *248–249*, 172–176. [CrossRef]
91. Gomes, M.P.; Juneau, P. Temperature and light modulation of herbicide toxicity on algal and cyanobacterial physiology. *Front. Environ. Sci.* **2017**, *5*, 50. [CrossRef]
92. Cobb, A.H.; Reade, J.P.H. Harmful algal blooms. In *Herbicides and Plant Physiology*; Cobb, A.H., Reade, J.P.H., Eds.; Wiley-Blackwell: Oxford, UK, 2010; pp. 176–197.
93. Gomes, M.P.; Le Manac'h, S.G.; Hénault-Ethier, L.; Labrecque, M.; Lucotte, M.; Juneau, P. Glyphosate-dependent inhibition of photosynthesis in willow. *Front. Plant Sci.* **2017**, *8*, 207. [CrossRef] [PubMed]

94. Gomes, M.P.; Le Manac'h, S.G.; Maccario, S.; Labrecque, M.; Lucotte, M.; Juneau, P. Differential effects of glyphosate and aminomethylphosphonic acid (AMPA) on photosynthesis and chlorophyll metabolism in willow plants. *Pestic. Biochem. Physiol.* **2016**, *130*, 65–70. [CrossRef] [PubMed]
95. Choi, C.J.; Berges, J.A.; Young, E.B. Rapid effects of diverse toxic water pollutants on chlorophyll a fluorescence: Variable responses among freshwater microalgae. *Weed Res.* **2012**, *46*, 2615–2626. [CrossRef] [PubMed]
96. Smedbol, É.; Lucotte, M.; Labrecque, M.; Lepage, L.; Juneau, P. Phytoplankton growth and PSII efficiency sensitivity to a glyphosate-based herbicide (Factor 540®). *Aquat. Toxicol.* **2017**, *192*, 265–273. [CrossRef] [PubMed]
97. Monsanto Europe S.A. Material Safety Data Sheet of Roundup Classic. Monsanto Europe S.A./N.V.: Antwerp, Belgium, 2015.
98. Syngenta. *Material Safety Data Sheet of Medallon Premium*; Syngenta Magyarország Kft.: Budapest, Hungary, 2018.
99. Scandinavian Culture Collection of Algae and Protozoa: Media Recipes. Available online: https://www.sccap.dk/media/ (accessed on 21 March 2024).
100. Allen, M.M. Simple conditions for growth of unicellular blue-green algae on plates. *J. Phycol.* **1968**, *4*, 1–4. [CrossRef] [PubMed]
101. *ISO 8692:2012*; Water Quality-Fresh Water Algal Growth Inhibition Test with Unicellular Green Algae. International Organization for Standardization: Geneva, Switzerland, 2012.
102. Organisation for Economic Co-operation and Development Test No. 201: Freshwater Alga and Cyanobacteria, Growth Inhibition Test. OECD Publishing, Paris. Available online: https://read.oecd-ilibrary.org/environment/test-no-201-alga-growth-inhibition-test_9789264069923-en#page1 (accessed on 21 March 2024).
103. Stevenson, R.J.; Lowe, R.L. Sampling and interpretation of algal patterns for water quality assessment. In *Rationale for Sampling and Interpretation of Ecological Data in the Assessment of Freshwater Ecosystems*; Isom, R.G., Ed.; American Society for Testing and Materials: Philadelphia, PA, USA, 1986; pp. 118–149.
104. McCormick, P.V.; Cairns, J. Algae as indicators of environmental change. *J. Appl. Phycol.* **1994**, *6*, 509–526. [CrossRef]
105. *ISO 10260:1992*; Water Quality-Measurement of Biochemical Parameters–Spectrometric Determination of the Chlorophyll-a Concentration. International Organization for Standardization: Geneva, Switzerland, 1992.
106. Lázár, D.; Takács, E.; Mörtl, M.; Klátyik, S.; Barócsi, A.; Kocsányi, L.; Lenk, S.; Domján, L.; Szarvas, G.; Lengyel, E.; et al. Application of a fluorescence-based instrument prototype for chlorophyll measurements and its utility in an herbicide algal ecotoxicity assay. *Water* **2023**, *15*, 1866. [CrossRef]
107. Barócsi, A.; Lenk, S.; Kocsányi, L.; Buschmann, C. Excitation kinetics during induction of chlorophyll a fluorescence. *Photosynthetica* **2009**, *47*, 104–111. [CrossRef]
108. Lichtenthaler, H.K.; Buschmann, C.; Knapp, M. How to correctly determine the different chlorophyll fluorescence parameters and the chlorophyll fluorescence decrease ratio RFd of leaves with the PAM fluorometer. *Photosynthetica* **2005**, *43*, 379–393. [CrossRef]
109. Kalaji, H.M.; Schansker, G.; Brestic, M.; Bussotti, F.; Calatayud, A.; Ferroni, L.; Goltsev, V.; Guidi, L.; Jajoo, A.; Li, P.; et al. Frequently asked questions about chlorophyll fluorescence, the sequel. *Photosynth. Res.* **2017**, *132*, 13–66. [CrossRef] [PubMed]
110. Brain, P.; Cousens, R. An equation to describe dose responses where there is stimulation of growth at low doses. *Weed Res.* **1989**, *29*, 93–96. [CrossRef]
111. Ritz, C.; Baty, F.; Streibig, J.C.; Gerhard, D. Dose-response analysis using R. *PLoS ONE* **2015**, *10*, e0146021. [CrossRef] [PubMed]
112. Ritz, C.; Jensen, S.M.; Gerhard, D.; Streibig, J.C. A hormesis effect on lettuce growth. In *Dose-Response Analysis Using R*; Ritz, C., Jensen, S.M., Gerhard, D., Streibig, J.C., Eds.; CRC Press: Boca Raton, FL, USA, 2020; pp. 23–26.
113. Wong, P.K. Effect of 2,4-D, glyphosate and paraquat on growth, photosynthesis and chlorophyll-a synthesis of *Scenedesmus quadricauda* Berb 614. *Chemosphere* **2000**, *41*, 177–182. [CrossRef] [PubMed]
114. Schönherr, J. A mechanistic analysis of penetration of glyphosate salts across astomatous cuticular membranes. *Pest Manag. Sci.* **2022**, *58*, 343–351. [CrossRef] [PubMed]
115. Ewacha, M.V.A.; Goldsborough, L.G. The response of *Scenedesmus quadricauda* and *Selenastrum capricornutum* to glyphosate toxicity (Roundup® formulation) with cellular growth and chlorophyll-a synthesis as endpoints. *Proc. Manitoba's Undergrad. Sci. Eng. Res.* **2013**, *1*, 1–19.
116. Powell, H.A.; Kerby, N.W.; Rowell, P. Natural tolerance of cyanobacteria to the herbicide glyphosate. *New Phytol.* **1991**, *119*, 421–426. [CrossRef]
117. Sáenz, M.E.; Di Marzio, W. Ecotoxicity of herbicide glyphosate to four chlorophyceaen freshwater algae. *Limnetica* **2009**, *28*, 149–158. [CrossRef]
118. LISEC Alga, growth inhibition test. Effect of MON 2139 on the growth of Selenastrum capricornutum. In *Monsanto Unpublished Study XX-89-093*; LISEC, Study Centre for Ecology and Forestry: Bokrijk, Belgium, 1989.
119. Cedergreen, N.; Streibig, J.C. The toxicity of herbicides to non-target aquatic plants and algae: Assessment of predictive factors and hazard. *Pest Manag. Sci.* **2005**, *61*, 1152–1160. [CrossRef]
120. Pereira, J.; Antunes, S.C.; Castro, B.B.; Marques, C.R.; Goncalves, A.M.M.; Goncalves, F.; Pereira, R. Toxicity evaluation of three pesticides on non-target aquatic and soil organisms: Commercial formulation versus active ingredient. *Ecotoxicology* **2009**, *18*, 455–463. [CrossRef]
121. Lipok, J.; Studnik, H.; Gruyaert, S. The toxicity of Roundup 360 SL formulation and its main constituents: Glyphosate and isopropylamine towards non-target water photoautotrophs. *Ecotoxicol. Environ. Saf.* **2010**, *73*, 1681–1688. [CrossRef]
122. Steber, J.; Guhl, W.; Stelter, N.; Schroder, F.R. Alkyl polyglycosides -ecological evaluation of a new generation of nonionic surfactants. *Tenside Surfactants Deterg.* **1995**, *32*, 515–521. [CrossRef]

123. Gomes, M.P.; Juneau, P. Oxidative stress in duckweed (*Lemna minor* L.) induced by glyphosate: Is the mitochondrial electron transport chain a target of this herbicide? *Environ. Pollut.* **2016**, *218*, 402–409. [CrossRef]
124. Rasch, A.; Hunsche, M.; Mail, M.; Burkhardt, J.; Noga, G.; Pariyar, S. Agricultural adjuvants may impair leaf transpiration and photosynthetic activity. *Plant Physiol. Biochem.* **2018**, *132*, 229–237. [CrossRef]
125. Guedes, R.N.C.; Cutler, G.C. Insecticide induced hormesis and arthropod pest management. *Pest Manag. Sci.* **2014**, *70*, 690–697. [CrossRef] [PubMed]
126. Mattson, M.P. Hormesis defined. *Ageing Res. Rev.* **2008**, *7*, 1–7. [CrossRef] [PubMed]
127. Calabrese, E.J.; Bachmann, K.A.; Bailer, A.J.; Bolger, P.M.; Borak, J.; Cai, L.; Cedergreen, N.; Cherian, M.G.; Chiueh, C.C.; Clarkson, T.W.; et al. Biological stress response terminology: Integrating the concepts of adaptive response and preconditioning stress within a hormetic dose-response framework. *Toxicol. Appl. Pharmacol.* **2007**, *222*, 122–128. [CrossRef]
128. Kendig, E.L.; Le, H.H.; Belcher, S.M. Defining hormesis: Evaluation of a complex concentration response phenomenon. *Int. J. Toxicol.* **2010**, *29*, 235–246. [CrossRef]
129. Bakonyi, G.; Szabó, B.; Seres, A. A hormézis, mint ökotoxikológiai jelenség, különös tekintettel a növényvédelemre. *bioKontroll* **2017**, *2*, 47–53. (In Hungarian)
130. Calabrese, E.J. Overcompensation stimulation: A mechanism for hormetic effects. *Critical Rev. Toxicol.* **2001**, *31*, 425–470. [CrossRef]
131. Jager, T.; Barsi, A.; Ducrot, V. Hormesis on life-history traits: Is there such thing as a free lunch? *Ecotoxicology* **2013**, *22*, 263–270. [CrossRef] [PubMed]
132. Forbes, V.E. Is hormesis an evolutionary expectation? *Functional Ecol.* **2000**, *14*, 12–24. [CrossRef]
133. Saxton, M.A.; Morrow, E.A.; Bourbonniere, R.A.; Wilhelm, S.W. Glyphosate influence on phytoplankton community structure in Lake Erie. *J. Great Lakes Res.* **2011**, *37*, 683–690. [CrossRef]
134. Vera, M.S.; Lagomarsino, L.; Sylvester, M.; Pérez, G.L.; Rodriguez, P.; Mugni, H.; Sinistro, R.; Ferraro, M.; Bonetto, C.; Zagares, H.; et al. New evidences of Roundup® (glyphosate formulation) impact on the periphyton community and the water quality of freshwater ecosystems. *Ecotoxicology* **2010**, *19*, 710–721. [CrossRef] [PubMed]
135. Jalal, A.; de Oliveira Junior, J.C.; Ribeiro, J.S.; Fernandes, G.C.; Mariano, G.G.; Trinidade, V.D.R.; dos Reis, A.R. Hormesis in plants: Physiological and biochemical responses. *Ecotoxicol. Environ. Saf.* **2021**, *207*, 111225. [CrossRef]
136. Panizzi, S.; Suciu, N.A.; Trevisan, M. Combined ecotoxicological risk assessment in the frame of European authorization of pesticides. *Sci. Total Environ.* **2017**, *580*, 136–146. [CrossRef]
137. Lozano, V.L.; Pizarro, H.N. Glyphosate lessons: Is biodegradation of pesticides a harmless process for biodiversity? *Environ. Sci. Eur.* **2024**, *36*, 55. [CrossRef]
138. Aparicio, V.C.; de Gerónimo, E.; Marino, D.; Primost, J.; Carriquiriborde, P.; Costa, J.L. Environmental fate of glyphosate and aminomethylphosphonic acid in surface waters and soil of agricultural basins. *Chemosphere* **2013**, *93*, 1866–1873. [CrossRef]
139. Brausch, J.M.; Smith, P.N. Toxicity of three POEA surfactant formulations to the fairy shrimp *Thamnocephalus platyurus*. *Arch. Environ. Contam. Toxicol.* **2007**, *52*, 217–222. [CrossRef]
140. Li, M.; Du, F.; Cao, C.; Li, B.; Zhai, X. Effect of glyphosate isopropylamine on the surface tension and surface dilational rheology properties of polyoxyethylene tallow amine surfactant. *J. Dispers. Sci. Technol.* **2016**, *37*, 213–221. [CrossRef]
141. Mesnage, R.; Antoniou, M.N. Ignoring adjuvant toxicity falsifies the safety profile of commercial pesticide. *Front. Public Health* **2018**, *5*, 361. [CrossRef] [PubMed]
142. Klátyik, S.; Takács, E.; Mörtl, M.; Földi, A.; Trábert, Z.; Ács, É.; Darvas, B.; Székács, A. Dissipation of the herbicide active ingredient glyphosate in natural water samples in the presence of biofilms. *Int. J. Environ. Anal. Chem.* **2017**, *97*, 901–921. [CrossRef]
143. European Commission. Commission regulation (EU) No 284/2013 of 1 March 2013 setting out the data requirements for plant protection products, in accordance with Regulation (EC) No 1107/2009 of the European Parliament and of the Council concerning the placing of plant protection products on the market. *OJ EU* **2013**, *L93*, 85–151. Available online: https://eur-lex.europa.eu/legal-content/EN/TXT/?uri=CELEX:32013R0284&qid=1707814054061 (accessed on 21 March 2024).
144. Seralini, G.-E. Pesticides in formulations: New revolutionary findings. *Toxics* **2024**, *12*, 151. [CrossRef] [PubMed]
145. Edwards, W.M.; Triplett, G.B.; Kramer, R.M. A watershed study of glyphosate transport in runoff. *J. Environ. Qual.* **1980**, *9*, 661–665. [CrossRef]
146. Servizi, J.A.; Gordon, R.W.; Martens, D.W. Acute toxicity of Garlon 4 and Roundup herbicides to salmon, Daphnia, and Trout. *Bull. Environ. Contam. Toxicol.* **1987**, *39*, 15–22. [CrossRef]
147. Mensink, H.; Janssen, P. Glyphosate. 1994. Available online: http://www.inchem.org/documents/ehc/ehc/ehc159.htm (accessed on 9 February 2024).

toxics

MDPI

Article

Ecotoxicological Evaluation of Safener and Antimicrobial Additives in Isoxaflutole-Based Herbicide Formulations

Eszter Takács [1,*], Diána Lázár [2,3], Augustine Siakwa [1], Szandra Klátyik [1], Mária Mörtl [1], László Kocsányi [4], Attila Barócsi [4], Sándor Lenk [4], Edina Lengyel [2,5] and András Székács [1]

1 Agro-Environmental Research Centre, Institute of Environmental Sciences, Hungarian University of Agriculture and Life Sciences, Páter Károly u. 1., H-2100 Gödöllő, Hungary; siakwaa@yahoo.com (A.S.); klatyik.szandra@uni-mate.hu (S.K.); mortl.maria@uni-mate.hu (M.M.); szekacs.andras@uni-mate.hu (A.S.)
2 Limnology Research Group, Center of Natural Science, University of Pannonia, Egyetem u. 10., H-8200 Veszprém, Hungary; lazar.diana@blki.hun-ren.hu (D.L.); lengyel.edina@mk.uni-pannon.hu (E.L.)
3 Aquatic Botany and Microbial Ecology Research Group, HUN-REN-BLKI, Klebelsberg Kuno u. 3, H-8237 Tihany, Hungary
4 Department of Atomic Physics, Institute of Physics, Budapest University of Technology and Economics, Műegyetem rkp. 3., H-1111 Budapest, Hungary; kocsanyi.laszlo@ttk.bme.hu (L.K.); barocsi.attila@ttk.bme.hu (A.B.); lenk.sandor@ttk.bme.hu (S.L.)
5 Limnoecology Research Group, ELKH-PE, Egyetem u. 10, H-8200 Veszprém, Hungary
* Correspondence: takacs.eszter84@uni-mate.hu; Tel.: +36-28522000 (ext. 1839)

Abstract: The environmental load by isoxaflutole and its formulated herbicide products has increasingly become apparent because, after the ban of atrazine, isoxaflutole has become its replacement active ingredient (a.i.). Obtaining information regarding the fate of this a.i. in environmental matrices and its ecotoxicological effects on aquatic organisms is essential for the risk assessment of the herbicide. In this study, the effects of Merlin Flexx- and Merlin WG75 formulated isoxaflutole-based herbicide products and two selected additives (cyprosulfamide safener and 1,2-benzisothiazol-3(2H)-one antimicrobial agent) were investigated on *Raphidocelis subcapitata* in growth inhibition assays. In ecotoxicological tests, two conventional (optical density and chlorophyll-a content) and two induced fluorescence-based (Fv*/Fp: efficiency of the photosystem PSII and Rfd* changes in the observed ratio of fluorescence decrease) endpoints were determined by UV-spectrophotometer and by our FluoroMeter Module, respectively. Furthermore, dissipation of isoxaflutole alone and in its formulated products was examined by an HPLC-UV method. In ecotoxicological assays, the fluorescence-based Rfd* was observed as the most sensitive endpoint. In this study, the effects of the safener cyprosulfamide and the antimicrobial agent 1,2-benzisothiazol-3(2H)-one on *R. subcapitata* is firstly reported. The results indicated that the isoxaflutole-equivalent toxicity of the mixture of the isoxaflutole–safener–antimicrobial agent triggered lower toxicity (EC_{50} = 2.81 ± 0.22 mg/L) compared to the individual effect of the a.i. (EC_{50} = 0.02 ± 0.00 mg/L). The Merlin Flexx formulation (EC_{50} = 27.04 ± 1.41 mg/L) was found to be approximately 50-fold less toxic than Merlin WG75, which can be explained by the different chemical characteristics and quantity of additives in them. The additives influenced the dissipation of the a.i. in Z8 medium, as the DT_{50} value decreased by approximately 1.2- and 3.5-fold under light and dark conditions, respectively.

Keywords: isoxaflutole; Merlin Flexx; Merlin WG75; cyprosulfamide; 1,2-benzisothiazol 3(2H)-one; safener; combined effect; herbicide; ecotoxicology; fluorescence

1. Introduction

Agricultural micropollutants (pesticide active ingredients, co-formulants, and other additives, mycotoxins, and fertilizers) can leach out from soil and contaminate surface water and drinking water supplies, and thus exert direct and indirect adverse effects on aquatic organisms and human health [1–5]. Agriculture, expanding in an effort to provide food for the growing and increasingly consuming human population, is forecasted to dramatically contribute to biodiversity collapse and become a major driver of global environmental change by 2050 [6–9]. Deterioration of water quality is also a key problem related to these contaminants [1,3,10,11], and consequently, internationally harmonized water quality indicators are in use to describe the state of water quality in agricultural areas and define the contribution of nutrient and pesticide pollution originating from agricultural activities [12].

Pesticide products exert their targeted activity by their active ingredients (a.i.s) interfering with key physiological processes in the pest organisms. In addition to the a.i.s, these products also contain various additives that can facilitate the stability, distribution, or efficacy of the a.i. or modify its adhesivity or other physicochemical characteristics. Three groups of additives need to be mentioned for the purpose of this study. Surfactants reduce the surface tension of the aqueous solution of the a.i., thus facilitating the spreading of the formulation used in plant protection as a film layer on the surface, thus aiding the absorption of the a.i. [13]. Antidotes (safeners) reduce or eliminate the phytotoxic effect of the a.i. on certain plants mostly by enhancing metabolic enzymes in these plants that can rapidly degrade the a.i. of pesticide formulations. The use of antidotes has several advantages in agricultural weed control: (i) selective eradication of weeds in botanically related crops, (ii) the use of non-selective herbicides for selective weed control, and (iii) the use of persistent soil herbicides by enhancing their degradation. An important drawback, however, is that the antidote may facilitate the emergence of weed resistance, as increased herbicide metabolism is a key mechanism in the development of not target-site-based weed resistance [14]. Antimicrobial and/or antifungal additives (preservatives) protect the a.i. from microbial degradation or unexpected chemical transformation on the plant surface, thus promoting enhanced efficacy of the formulation [15]. As seen, antidotes (safeners) and antimicrobial additives are biologically active components. Therefore, they are not typical additives, as additives are, in principle, inert regarding the pesticidal mode of action of the a.i., and therefore, formulation substances used to be evaluated as inert components in the regulatory risk assessment process. Certain components in pesticide formulations applied to animals, including the sodium salts of sulfomethylated lignosulfonic acid suspension/emulsion stabilizers, are to date exempt from the requirement of tolerance in the U.S. as "inert ingredients" [16]. Nonetheless, numerous studies confirmed that the toxicity of formulated pesticides is often higher than that of the a.i.(s) themselves, confirming that the formulation agents may exert their own toxicity on non-target organisms, possibly additive or in some cases synergistic with the a.i.s [17].

Isoxaflutole (5-cyclopropylisoxazol-4-yl-2-mesyl-4-trifluoromethylphenyl-ketone), developed in 1995, is the a.i. of several commercial selective systemic herbicides for pre-emergent weed control [18]. The mode of action of isoxaflutole is the inhibition of the enzyme 4-hydroxyphenylpyruvate dioxygenase (HPPD). As a pigment inhibitor, it inhibits the biosynthesis of carotenoid pigments that protect chlorophyll from degradation by sunlight. Excessive exposure to sunlight causes photo-oxidation of chlorophyll pigments, degradation of chloroplasts, and consequent whole-plant death [19–21]. The molecular feature in isoxaflutole responsible for binding to HPPD has been identified to be the diketonitrile moiety formed in situ, and more potent derivatives were reported by including two of such moieties in a single molecule [22]. Nonetheless, isoxaflutole remains the sole isoxazole derivative among commercialized HPPD inhibitor herbicides, often used in herbicide-resistant weed management strategies [23–25]. Isoxaflutole was first registered for use on maize in 1999, and it is currently authorized in the European Union until 2034 (assessed by Sweden as rapporteur Member State) [26,27]. It was also registered in the US for use on genetically modified (GM) isoxaflutole-tolerant soybeans in 2020 [28–30]. With the introduction of GM crops into agricultural practice, the rate and amount of the a.i. applied increases, as shown by the worldwide use of glyphosate following the approval of tolerant GM crops [31,32]. Nonetheless, the adoption of isoxaflutole-resistant GM soy remains limited due to restrictions in the use of isoxaflutole [25].

1.1. Occurrence of Isoxaflutole and Its Additives in the Environment

Isoxaflutole has been detected as an emerging contaminant in surface and drinking waters due to its increasing use, together with the metabolites of diketonitrile, dichloroacetonitrile, and benzoic acid [33,34]. It has been identified as a persistent pollutant in water [21] exerting algal toxicity [35], but long-term accumulation of it and its main diketonitrile metabolite was not evidenced in semistatic water bodies [36]. Its half-life (DT_{50}) water is 18 days [37] and can be substantially shorter (0.5–14 days) in soil [38] due to its leaching potential via migration through the unsaturated zone [39]. Due to the known hydrolytic decomposition of isoxaflutole and subsequent conversion to cyto- and genotoxic dichloroacetonitrile [34], the adverse biological effects of isoxaflutole can also be attributed to this latter metabolite. Due to the increasing application rates of isoxaflutole accelerated by the cultivation of isoxaflutole-resistant GM soy, its chemical load on the environment increases, yet environmental accumulation risk has not been attributed to the a.i. In cases of severe environmental pollution with pesticide a.i.s causing persistent or cumulating contamination, not only the environmental fate and toxicology of the substance, but potential means of remediation are needed to be assessed, as seen, e.g., in the case of paraquat [40] or glyphosate [32,41,42]. The environmental status of isoxaflutole is fortunately not as unfavorable, apparently manageable by certain application restrictions, and therefore the feasibility of remediation is uncalled for.

Cyprosulfamide (N-[[4-[(cyclopropylamino)carbonyl]phenyl]sulfonyl]-2-methoxyben zamide), developed in 2009, is a herbicide safener used with herbicide a.i.s including isoxaflutole. It is detected along with its metabolites in surface water, but not in ground water [43,44], and was found toxic but unlikely to cause lethality but exert adverse effects upon chronic exposure on *Daphnia magna* at relevant environmental concentrations [44]. As a common additive to frequently used pesticide formulations, cyprosulfamide and its two metabolites, its desmethyl derivative and N-cyclopropyl-4-sulfamoylbenzamide, were detected in up to 56% of the 34 surface water (but not groundwater) samples collected near cornfields in the midwestern United States. Thus, cyprosulfamide, cyprosulfamide desmethyl, and N-cyclopropyl-4-sulfamoylbenzamide were found in 25%, 19%, and 56% frequency with highest concentrations and detection rates during the growing season, and with maximum concentrations ranging between 22.0 and 5185.9 ng/L [43].

1,2-Benzisothiazoline-3-one is a preservative, disinfectant, and industrial biocide and antimicrobial agent in industrial and consumer products including formulated pesticides. Due to its widespread use, it has been detected as an aquatic environmental contaminant [45–47].

1.2. Algae as Indicators of Water Quality

Algal biomass, as an indicator of water quality and ecotoxicological impacts, can be measured using a variety of techniques. The most commonly used methods include the determination of algal cell numbers by microscope, optical density (OD), and the chlorophyll content of algal cells after alcohol extraction by spectrophotometry, and dry mass measurements [48]. The primary source of endogenous fluorescence in algae is the fluorescence signal induced by chlorophylls responsible for photosynthesis [49]. Thus, the efficiency of photosynthesis can also be characterized by fluorescence induction kinetics describing changes in the photosynthetic process and the physiological state of algal cultures [50]. The monocultures of various microalgal species (e.g., *Raphidocelis subcapitata*) are often used as test organisms in ecotoxicological studies to determine the side effects of agricultural pollutants.

This study was targeted to determine the phytotoxic effects of the herbicide a.i. isoxaflutole, its commercial formulated herbicide products (Merlin Flexx and Merlin WG75), and selected additive components used in these products (cyprosulfamide, 1,2-benzisothiazol-3(2H)-one) on *R. subcapitata* as a water quality indicator algal species, using our previously developed, modular microplate-based fluorometer prototype [35] and conventional methods. The investigation also included the assessment of the aquatic stability of the a.i. alone and within the formulation. The aim of the study was to demonstrate the effects of the formulating agents on the environmental fate and the algal toxicity of the herbicide a.i. isoxaflutole.

2. Materials and Methods

2.1. Monoculture of *Raphidocelis Subcapitata*

The unicellular model green algae species *R. subcapitata*, Korshikov (NIVA-CHL1) were obtained from the alga collection of the Norwegian Institute for Water Research (NIVA). This microalga is commonly applied in ecotoxicological investigations as an indicator species due to its ubiquitous distribution and high sensitivity against environmental pollution. For maintenance of the batch culture of microalgae *R. subcapitata* and the preparation of different solutions of test compounds, Zehnder 8 (Z8) [51] media were used. The culture was maintained at 23 ± 1 °C and illuminated in a 14:10 light/dark period by cool-white fluorescence tubes with a photosynthetic photon flux density of 15 μmol/m^2/s. Fresh media were added to the cultures every two weeks.

2.2. Test Compounds

The effects of two isoxaflutole-based herbicide products, Merlin Flexx and Merlin WG75, were investigated. Merlin 75WG, in contrast with Merlin Flexx, is not authorized in Hungary, thus it was purchased from Slovakia. Isoxaflutole (CAS 141112-29-0), cyprosulfamide (CAS 221667-31-8), and 1,2-benzisothiazol-3(2H)-one (CAS 2634-33-5) were obtained from Merck and were of ≥97.5%, ≥98.0%, and ≥96.5%, respectively.

There appear characteristic differences between the two herbicide preparations regarding their physico-chemical properties and chemical composition. Merlin Flexx is an aqueous suspension concentrate, while Merlin WG75 is a water dispersible granulate. Although the a.i. is isoxaflutole in both preparations, the concentration of the a.i. and the chemical structure of additives applied during the formulating process are different (Table 1). In this study, the individual effects of the antidote cyprosulfamide and the antimicrobial agent 1,2-benzisothiazol-3(2H)-one; the combined effects of the a.i. and the antidote/antimicrobial agent; and the effects of the two formulated products were investigated. The combined effects of isoxaflutole + cyprosulfamide and isoxaflutole + cyprosulfamide + 1,2-benzisothiazol-3(2H)-one were investigated according to the concentration ratio in the Merlin Flexx formulated product (Table 1). As for 1,2-benzisothiazol-3(2H)-one, a concentration range (0.005–0.05%) is given by the supplier in Safety Data Sheet, and the compound was tested at 0.05% in the combination study. Information on the ingredients of the preparations is available from their safety data sheets according to the Regulation (EC) No. 1907/2006 [52–54].

Table 1. Composition of the isoxaflutole-based herbicide products.

Herbicide Products	Concentration (%)	CAS No [1]	Function in the Herbicide Product
Merlin Flexx			
isoxaflutole	20.3	141112-29-0	active ingredient
cyprosulfamide	20.3	221667-31-8	additive: antidote
1,2-benzisothiazol-3(2H)-one	0.005–0.05	2634-33-5	additive: antimicrobial
D-glucopyranose, oligomer, C9-11-alkyl glycosides	3–10	132778-08-6	additive: surfactant
glycerin	>1	56-81-5	humectant, antigelling, and antifreeze agent
Merlin WG75			
isoxaflutole	75	141112-29-0	active ingredient
sodium diisopropylnaphthalene sulphonate	3–10	1322-93-6	additive: surfactant
lignosulfonic acid, sodium salt, sulfomethylated	3–10	68512-34-5	suspension/emulsion stabilizer
kaolin	>1	1332-58-7	carrier
pyrogenic (fumed) amorphous silica	>1	112945-52-5	carrier

[1] Chemical Abstract Service Number.

2.3. Bioassay

Ecotoxicological algal growth inhibition tests were performed to investigate the possible harmful effect of the test compounds on the *R. subcapitata* monocultures according to the standardized OECD Guideline 201 [55]. The 72 h growth inhibition was described by the biomass reduction of the algal cultures, whereby optical density (OD) and chlorophyll-a (Chl-a) content were measured as endpoints. Chl-a content was detected after an extraction process that was performed according to the ISO 10260:1992 standard [56] and the Felföldy formula was applied to quantify the Chl-a content [57].

During the 72 h of tests, the following parameters were controlled: light intensity (continuous, $104.9 \pm 14.9\ \mu E/m^2/s$) by illumination (continuous, 2500 lux), temperature (22 ± 2 °C), pH of the algal Z8 medium (pH = 6–7), and intensity of stirring (continuous, 100 rpm). The ecotoxicological assays were performed in three replicates for each test substance diluted serially. In the three individual experiments, untreated control and alga suspensions exposed to different concentrations of the test substances were evaluated.

Besides the conventional parameters (OD and Chl-a content), two fluorescence-based endpoints were also measured in algal growth inhibition assays via induced fluorescence. Both the proxy of quantum efficiency of the algae photosystem PSII (F_v^*/F_p) and the changes in the observed ratio of fluorescence decrease (Rfd*) and describe the status of the photosynthetic activity of the plants. Both parameters are calculated from fluorescence quantities observed by the FMM instrument: F_o—non-variable fluorescence intensity; F_p—peak fluorescence intensity, maximum fluorescence at a non-saturating light pulse; F_s—steady-state (terminal) fluorescence. F_v^*/F_p and Rfd* were calculated according to the following equations: $F_v^* = F_p - F_o$; $F_d = F_p - F_s$; $Rfd^* = F_d/F_s$ [35]. For measurement of the fluorescent parameters, 250 μL treated or untreated algal suspension was added into the selected wells of a 96-well microplate. Circumstances (pH, temperature, light intensity, etc.) of the ecotoxicity assay were set up according to the OECD guideline; thus, measurements of OD, Chl-a content, F_v^*/F_p, and Rfd* were performed under the same conditions.

Growth inhibition by compounds tested in this study was investigated in two ways. Firstly, based on the equation of the dose–response curve EC_{50}, values were calculated. Secondly, according to the endpoint parameters, growth inhibition rate (IR) values were determined based on the following formula: IR = (C − T)/T*100, where C and T are the density of algal cells in the control and experimental group, respectively. IR values were calculated and compared at the isoxaflutole EC_{50} equivalent concentration in all mixtures tested. Concentration of combinations were expressed by the sum of the amount of compounds in the combination.

2.4. Analytical Determination of Isoxaflutole Substance Loss in Its Formulated Products

The dissipation of isoxaflutole was determined under dark and light conditions in distilled water and in Z8 medium applied to maintain the algal monocultures. Dark and light conditions were applied in parallel experiments using the same treatment periods. Dissipation of isoxaflutole was determined using an analytical standard of the a.i. and samples of the two herbicide formulations Merlin Flexx and Merlin WG75. The analytical determination of isoxaflutole was performed by high-performance liquid chromatography with ultraviolet detection (HPLC-UV). A mixture of acetonitrile and water (70:30) was used as an eluent at a flow rate of 1 mL/min. Isoxaflutole was separated on a PerfectSil 100 ODS-3 column (MZ-Analysentechnik GmbH, Mainz, Germany) (150 × 4.6 mm i.d., 5 μm) at 35 °C, and UV detector signals were recorded at $\lambda = 220$ nm. The concentration of the a.i. was determined with an initial isoxaflutole-equivalent concentration of 5 μg/mL in samples collected in triplicates at 0, 9, 24, 48, 72, 96, 120, 168, 216, 264, and 336 h of treatment.

2.5. Instrumentation

Ecotoxicological algal growth inhibition assays on monocultures of R. subcapitata were performed in a shaking incubator (Witeg WIS-10RL, Wertheim, Germany) at controlled parameters (see Section 2.3). Chl-a content and OD values of algal suspensions were measured by a spectrophotometer (UV/VIS Camspec single beam M330, Camspec, Crawley, UK). HPLC-UV determinations were carried out using a Youngin YL9100 HPLC instrument equipped with a YL9150 autosampler (Youngin Chromass, Anyang-si, Korea).

Fluorescent parameters $F_v{}^*/F_p$ and Rfd* were measured by the FluoroMeter Module (FMM) [35,58] equipped with an apparatus capable of holding standard-size 96-well microplates and allowing manual stepping among the wells, which is a modified version of a plant leaf fluorometer capable of measuring the excitation kinetics of Chl-a fluorescence induction besides the traditional Kautsky induction kinetics [59–62]. The main parameters of the instrument are summarized in Table 2. The limit of detection and the lower limit of quantification are 4.01×10^6 and 8.12×10^6 cells for *R. subcapitata*, respectively. The latter was used throughout the experiments and the kinetic curves were detected simultaneously at the two maxima of the Chl-a fluorescence (at the 690 nm red and 735 nm far-red bands) upon continuous excitation with no saturation pulses. Wavelengths applied in the determination of the OD, Chl-a content, $F_v{}^*/F_p$, and Rfd* parameters by spectrophotometry and the induced-fluorescence method are summarized in Table 3.

Table 2. Features of the FluoroMeter Module applied in ecotoxicity assays on *Raphidocelis subcapitata*.

FluoroMeter Module	Feature/Type	Role
actinic light source	635 nm laser diode with 256-step digital optical power adjustment	excitation
interference filters	NT43-089 for 690 nm and NT43-091 for 730 nm, full width at half maximum of 10 nm each [1]	separation of the detection wavelengths and eliminate the scattered illumination light
cut-off filters	RG665 for 665 nm [1]	
photodetector	Low-noise PIN photodetectors (SD-200-14-21-241) [2]	photocurrent-to-voltage conversion
electrometer amplifier	OPA129 [3]	
signal amplifier	AD620 [4]	
digitalizer	12-bit ADC (AD7864-2) [4]	signal digitalization
computer	single-board computer (CMD16686GX) [5]	firmware, storage of data

[1] Edmund Optics, Barrington, NJ, USA; [2] Laser Components, Olching, Germany; [3] Texas Instruments, Dallas, TX, USA; [4] Analog Devices, Willington, MA, USA; [5] Real Time Devices, Wickwar, UK.

Table 3. Wavelengths applied in determination of endpoint parameters of ecotoxicological algal growth inhibition assays.

Method	Parameter Measured	Wavelength [1]
spectrophotometry	Optical density	detection: 750 nm
spectrophotometry	Chl-a content	detection: 750 nm [2], 666 nm [3], 653 nm [4]
induced-fluorescence	$F_v{}^*/F_p$	excitation: 635 nm detection: 690 nm, 735 nm
induced-fluorescence	Rfd*	excitation: 635 nm detection: 690 nm, 735 nm

[1] Applied in the measurement procedure of the given parameter. [2] Degree of turbidity. [3] First detection wavelength of Chl-a. [4] Second detection wavelength of Chl-a.

2.6. Statistical Evaluation

Effects of tested compounds on *R. subcapitata* were described by $EC_{50} \pm SD$ mg/L values for the two formulated herbicide products (Merlin Flexx and Merlin WG75), cyprosulfamide safener, and 1,2-benzisothiazol-3(2H)-one antimicrobial agent. The results of the ecotoxicity assays were statistically analyzed by the statistical software R 4.0 (The R Foundation for Statistical Computing, Vienna, Austria). Differences between the EC_{50} values (calculated on the basis of the dose–response equation) for OD, Chl-a content, and

Rfd* were detected by general linear models at a 5% significance level. Shapiro–Wilk and Levene's or Bartlett's test at a significance level of 5% were applied for the determination of normality and homogeneity of variance, respectively. The applicability of the fitted model was checked in each case with diagnostic plots (QQ plot, residual variances, Cook's distance plot). Tukey's honest significant difference (HSD) tests were performed as post hoc analyses to assess the significant differences between groups.

3. Results

3.1. Ecotoxicological Evaluation of Isoxaflutole-Based Herbicide Formulations

Ecotoxicity studies were performed to determine the effect of isoxaflutole-based formulated herbicide products (Merlin Flexx and Merlin WG75) and two additives applied in the formulating process of Merlin Flexx: cyprosulfamide safener and 1,2-benzisothiazol-3(2H)-one antimicrobial agent. The effects of these compounds were tested on *R. subcapitata* in growth inhibition assays based on the respective OECD guideline. Table 4 summarizes the highest concentration of test compounds investigated and the growth inhibition rate at these concentrations from three independent experiments. The highest concentrations were chosen to obtain a data point at the upper plateau of the sigmoid dose–response curve (concentration vs. growth inhibition). The lower concentrations investigated were sequentially diluted with a dilution factor of 1:2.

Table 4. Growth inhibition rates at the highest concentrations investigated in this study.

Substance	The Highest Concentration Investigated in This Study (mg/L)	IR (%)
active ingredient		
isoxaflutole	0.064	98.0 ± 1.0
herbicide products		
Merlin Flexx	80	96.3 ± 0.6
Merlin WG75	1.3	96.0 ± 4.7
combinations		
cyprosulfamide	100	21.3 ± 1.5
1,2-benzisothiazol-3(2H)-one	0.6	
isoxaflutole + cyprosulfamide	10 + 10	96.3 ± 1.5
isoxaflutole + cyprosulfamide + 1,2-benzisothiazol-3(2H)-one	5 + 5 + 0.025	95.7 ± 3.0

Growth inhibition was determined by measuring OD, Chl-a content, and two induced fluorescence-based parameters (F_v*/F_p and Rfd*) characterizing the status of photochemical systems of the microalgal indicator species *R. subcapitata*. Concentrations resulting in 50% growth inhibition are presented in Table 5. Algal toxicity of the a.i. isoxaflutole is listed as reported in our previous study [35].

For all the test compounds and their combinations tested, significantly lower EC_{50} values (p: 0.005–0.030) were determined based on Chl-a content than based on the OD value. The toxicity order of the test compounds is the same for both ecotoxicological endpoints. The most toxic ingredient was found to be the a.i. isoxaflutole ($EC_{50(OD)}$ = 0.03 ± 0.00 mg/L and $EC_{50(Chl\text{-}a)}$ = 0.02 ± 0.00 mg/L). In the case of Merlin Flexx, the additives (see Table 1) reduced the toxicity of the a.i. by ~200-fold, as the isoxaflutole equivalent EC_{50} was found to be 6.80 ± 0.02 mg/L mg/L. The a.i. equivalent concentration is the concentration of the a.i. in the formulated herbicide product at which it exerts a 50% effect. An $EC_{50(OD)}$ of 0.74 ± 0.20 mg/L was determined for Merlin WG75, indicating that the toxicity of isoxaflutole is substantially reduced (by one order of magnitude) by the additives present (e.g., kaolin).

Table 5. Ecotoxicological effects of the active ingredient isoxaflutole and its formulations on the growth of the algal species *Raphidocelis subcapitata* as determined by optical density (OD), Chl-a content, and Rfd*.

Substance	Method of Detection [1]	$EC_{50} \pm SD$ (mg/L)	Isoxaflutole-Equivalent $EC_{50} \pm SD$ (mg/L) [2]
active ingredient			
isoxaflutole [3]	OD	0.03 ± 0.00	0.03 ± 0.00
	Chl-a	0.02 ± 0.00	0.02 ± 0.00
	Rfd*	0.02 ± 0.00	0.02 ± 0.00
herbicide products			
Merlin Flexx	OD	33.3 ± 1.10	6.80 ± 0.20
	Chl-a	28.52 ± 0.32	5.84 ± 0.32
	Rfd*	27.04 ± 1.41	5.48 ± 0.48
Merlin WG75	OD	0.74 ± 0.20	0.55 ± 0.17
	Chl-a	0.64 ± 0.10	0.50 ± 0.07
	Rfd*	0.58 ± 0.15	0.43 ± 0.11
additives			
cyprosulfamide	OD	>100	–
	Chl-a	>100	–
	Rfd*	>100	–
1,2-benzisothiazol-3(2H)-one	OD	0.20 ± 0.00	–
	Chl-a	0.10 ± 0.01	–
	Rfd*	0.18 ± 0.01	–
combinations			
isoxaflutole + cyprosulfamide	OD	1.37 ± 0.04	0.69 ± 0.02
	Chl-a	1.22 ± 0.06	0.61 ± 0.03
	Rfd*	0.91 ± 0.07	0.46 ± 0.04
isoxaflutole + cyprosulfamide + 1,2-benzisothiazol-3(2H)-one	OD	7.43 ± 0.58	3.71 ± 0.29
	Chl-a	6.33 ± 0.44	3.12 ± 0.22
	Rfd*	5.64 ± 0.45	2.81 ± 0.22

[1] Methods of detection of algal biomass growth of *R. subcapitata* determined by optical density (OD), by extracted Chl-a content, and by the observed ratio of fluorescence decrease of the algal photosystem PSII (Rfd*) detected by our modular microplate-based fluorometer prototype [35]. [2] The 72 h EC_{50} value for the active ingredient (a.i.) represents the concentration of the a.i. present in the mixture or product that exerts a 50% effect in the given biotest. [3] Published earlier by Lázár et al., 2023 [35].

For 1,2-benzisothiazol-3(2H)-one, an $EC_{50(OD)}$ of 0.20 ± 0.00 mg/L was determined, whereas cyprosulfamide used as an antidote for isoxaflutole did not show any algal toxicity below 100 mg/L, which is the so-called limit test concentration as defined in the corresponding OECD guideline (Table 1) [55]. EC_{50} values indicated that the presence of cyprosulfamide and cyprosulfamide + 1,2-benzisothiazol-3(2H)-one resulted in ~23-fold and ~123-fold lower toxic effects of isoxaflutole, respectively, based on the OD values detected.

The determination of two photochemical parameters (F_v*/F_p—photochemical efficiency of the PSII photochemical system and Rfd*—fluorescence decrease ratio) was performed using the FMM fluorescence-based instrument. Both parameters characterize the functioning of the PSII photochemical system. The parameter F_v*/F_p was not found to be a suitable endpoint for the study of the effect of isoxaflutole and its formulations on green microalgae, showing no concentration dependence between treatments. However, the EC_{50} values based on the Rfd* parameter were determined to be 0.02 ± 0.0 mg/L, 27.04 ± 1.41 mg/L, 0.58 ± 0.15 mg/L, >100 mg/L, and 0.18 ± 0.01 for isoxaflutole, Merlin Flexx, Merlin WG75, cyprosulfamide and 1,2-benzisothiazol-3(2H)-one, respectively. The results were significantly lower (p: 0.009–0.040) for all the compounds and their mixtures tested compared to EC_{50} values determined based on OD and Chl-a content (the difference was not significant in the case of Merlin Flexx and isoxaflutole compared to Chl-a content), suggesting that the vitality index determined by induced fluorescence is a more sensitive parameter compared to the determination of OD and Chl-a content after alcohol extraction. It also provides additional information on the effect of the test substance. Growth inhibition rates calculated for isoxaflutole EC_{50} values equivalent for herbicide products and combinations of the a.i. and additives are presented in Table 6.

Table 6. Growth inhibition rate of the herbicide products and combinations at isoxaflutole EC_{50} equivalent concentrations.

Substance	Isoxaflutole EC_{50} Equivalent Concentration (mg/L)	IR (%)	
active ingredient			
isoxaflutole	0.03	OD	50.0
	0.02	Chl-a	50.0
	0.02	Rfd*	50.0
herbicide products			
Merlin Flexx	0.15	OD	0.22
	0.1	Chl-a	0.17
	0.1	Rfd*	0.18
Merlin WG75	0.04	OD	3.64
	0.03	Chl-a	3.00
	0.03	Rfd*	3.49
combinations			
isoxaflutole + cyprosulfamide	0.06	OD	2.17
	0.04	Chl-a	1.64
	0.04	Rfd*	2.17
isoxaflutole + cyprosulfamide + 1,2-benzisothiazol-3(2H)-one	0.065	OD	0.88
	0.045	Chl-a	0.72
	0.045	Rfd*	0.8

3.2. *Analytical Determination of Substance Loss of Isoxaflutole*

Isoxaflutole, as the replacement compound for atrazine, has been detected as an emerging water pollutant in the US and the EU [33,63]. For environmental risk assessment, it is crucial to obtain information regarding the fate of pollutants in different environmental matrices. The aim of this analytical determination was to ensure the accuracy of algal toxicity tests, the stability of the compound during the test, and to gather information about the dissipation time (DT_{50}) value of isoxaflutole in the presence of additives applied in the herbicide formulations. The dissipation of isoxaflutole in the form of an analytical standard and also in its formulated products was determined in distilled water and Z8 medium, under dark and light conditions with initial isoxaflutole and isoxaflutole-equivalent concentrations of 5 μg/L. The results showed no apparent dissipation of isoxaflutole in distilled water within 2 weeks. For Z8 media, the results are presented in Figure 1. In the two formulations, the analytical determination of the samples has not immediately reached the initial concentration of 5 μg/L at 0 h due to a slight time delay of micelle formation of isoxaflutole with the surfactant additives in the formulation. Considerable differences appeared regarding the dissipation of isoxaflutole in pure form and formulated herbicides both under dark and light conditions. Dissipation of the a.i. isoxaflutole as an analytical standard was completed on the 216th hour (DT_{50} = 131 h) and the 120th (DT_{50} = 59 h) under light and dark conditions, respectively. In both conditions, the additives in the formulations Merlin Flexx and Merlin WG75 (Table 1) contributed to a longer dissipation process. Under light conditions, the substance loss was completed on the 336th hour for both formulations, with DT_{50} values of 154 h and 158 h for Merlin Flexx and Merlin WG75, respectively. In contrast, under dark conditions, the dissipation remained incomplete during the experiment. DT_{50} values were found to be 194 h and 222 h for Merlin Flexx and Merlin WG75, respectively.

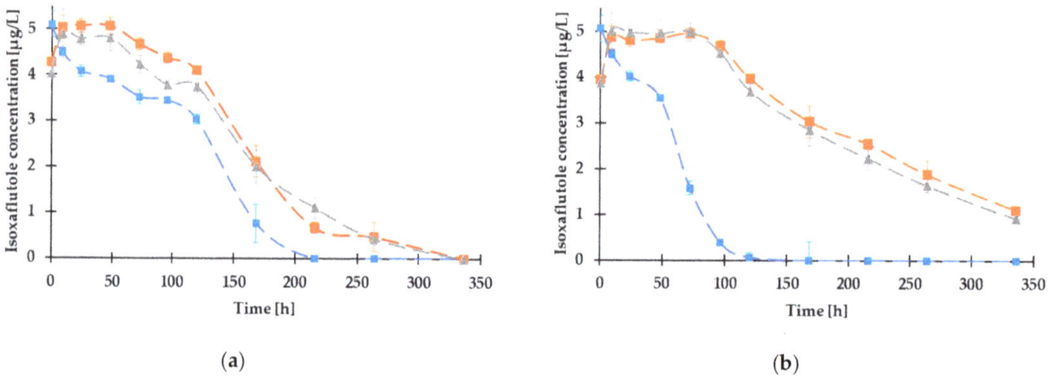

Figure 1. Dissipation of isoxaflutole in the form of an analytical standard (blue lines) and formulated herbicide products Merlin Flexx (grey lines) and Merlin WG75 (orange lines) under (**a**) light and (**b**) dark conditions. Data for isoxaflutole in form of an analytical standard under light conditions were published earlier [35].

The additives in formulated herbicides had a considerable influence on the dissipation of isoxaflutole, as they delayed the dissipation time and resulted in higher DT_{50} values. However, results indicated that the differences between dark and light circumstances were more remarkable than the chemical characteristics of the additives in the two formulated products.

4. Discussion

The mode of action of isoxaflutole is the inhibition of the enzyme 4-hydroxyphenylpyruvate dioxygenase, thus the quite high toxicity of the a.i. to the algal species *R. subcapitata* is not surprising, as it acts on all green plant organisms. Isoxaflutole has been shown to exert phytotoxicity on green algae: its 72 h EC_{50} values have been reported to range between 0.003 and 0.380 mg/L on freshwater green or diatom algae and duckweed [64], 0.030–1.71 mg/L on *R. subcapitata* [18,35,65] with 0.12 mg/L on 120 h exposure time [66], and 0.016–0.140 mg/L on *Selenastrum capricornutum* [18,64]. Similar inhibitory effects of isoxaflutole have been demonstrated in a plant spectral analysis system for herbicide efficacy using a terrestrial model weed crabgrass (*Digitaria ciliaris*) as a test plant and chlorophyll fluorescence as an endpoint [67]. The additional 96-well microplate-based biotest detected the effects on chlorophyll fluorescence, but instead of applying fluorimetry as in our case, it used a semiconductor device, a complementary metal oxide semiconductor camera. However, wide variability is seen among aquatic plant organisms regarding their sensitivity to this a.i. [68]. In our present study, a reduced toxicity of the two formulated preparations compared to the a.i. was observed, due to the combined effect of the a.i. and the additives.

Although it has already been reported for various pesticide a.i.s that the presence of additives in pesticide formulations can alter the toxicity of the a.i. [69–73], our results are the first that demonstrate the same phenomenon for isoxalfutole-based formulations. Moreover, the individual toxicity of numerous additives has also been published [74–76], among which the first generation of polyethoxylated amines has been the focus recently. As a result of numerous independent studies [74,77–79], this surfactant-type was banned in 2016 as a co-formulant in the European Union and was progressively replaced by other surfactants, ethoxylated etheramines, which exhibited lower toxicity on non-target organisms [80,81]. For both, the composition of the preparations and the physico-chemical characteristics of the additives were different, which exerted significantly different toxic effects. Thus, the influence of additives applied in formulated products on the toxic properties of the herbicide formulations is evident.

Nonetheless, the views on the health safety of isoxaflutole have not been unambiguously positive. As early as during its initial assessment and subsequent toxicological evaluation, the US EPA classified the carcinogenic potential of isoxaflutole in 1997 as "likely" to be carcinogenic to humans by all routes of exposure [82], and stated later in 2011: "In carcinogenicity studies, isoxaflutole induced liver and thyroid tumors in rats and liver tumors in mice. Isoxaflutole was classified as "likely to be a human carcinogen". The method of quantification was linear cancer slope factor (Q1*)." [83]. The European Food Safety Authority (EFSA) proposed to classify it as a carcinogen category 2 (suspected human carcinogen) due to observed thyroid follicular cell adenoma in rats and liver carcinoma in mice [65]. These tumor-inducing effects of isoxaflutole are suggested to lay on the grounds of endocrine disruption [84]. Thyroid tumor induction was considered to occur via a possible disruption of the thyroid–pituitary hormonal feedback mechanisms [82], and since this proposed mode of thyroid tumor induction can apply to humans as well, a warning has recently been published that this mode of action of isoxaflutole is of public health relevance [85]. However, approval of the a.i. was approved as no experimental evidence proved the assumption of the carcinogenicity of isoxaflutole on humans.

Information on the effects of both formulations regarding the toxicity to aquatic plant is only available in their MSDS [52,53]. For Merlin Flex and Merlin WG75, EC_{50} values are described for *R. subcapitata* and *Desmodesmus subspicatus* green algal species, respectively. Varying by the endpoints detected in this study, a 4.1–5.1-fold lower toxicity for Merlin Flexx, and a minimum 46.9–59.8-fold higher toxicity for Merlin WG75 on *R. subcapitata* were found compared to data in the MSDSs. The difference in the case of Merlin WG75 is due to the different sensitivities of different algal species.

The antimicrobial additive 1,2-benzisothiazol-3(2H)-one in Merlin Flexx is highly toxic to aquatic ecosystems and its presence in surface waters has been demonstrated [45–47]. Wang et al. studied the effect of the additive on three species of green algae. For *Scenedesmus* sp. LX1, *Chlorella* sp. HQ, and *Chlamydomonas reinhardtii*, the detected EC_{50} values were 1.70, 0.41, and 1.16 mg/L, respectively, indicating the differences in sensitivity between species and the importance of including more species in the environmental risk assessment. It was found that the primary effect of 1,2-benzisothiazol-3(2H)-one was to inhibit photosynthesis. This inhibitory effect appeared reversible at exposure in the concentration range of 1–30 mg/L [86]. In our study, *R. subcapitata* appeared to be 2–8-fold more sensitive to the additive (EC_{50} = 0.20 ± 0.00 mg/L) than the above-mentioned algal species. The cyprosulfamide additive did not show any toxic effect at a concentration of 100 mg/L on the algae species tested. As an antidote, it protects the plant against the toxic effect of the a.i. isoxaflutole, an effect which was presumably also observed in *R. subcapitata*. Its lower toxicity compared to Merlin Flexx is probably due to this mechanism of action. A similar moderating effect of another safener benoxacor on the toxicity of the herbicide a.i. iodosulfuron–methyl–sodium was reported on *Desmodesmus subspicatus* [87].

Our dissipation study of isoxaflutole in the forms of analytical standard and formulated products indicated that the presence of additives in the formulation prolonged the DT_{50} value of isoxaflutole in water. In contrast to the ecotoxicity experiment on *R. subcapitata* alga species, in the dissipation study, the composition of the formulated products and the physico-chemical characteristics of the additives did not significantly influence the dissipation time. The alteration of dissipation time when the a.i. and the additive are present in water samples at the same time has been also published for neonicotinoid-type and glyphosate a.i.s. in surface water [77,88,89] and for esfenvalerate in seawater samples [90], and slight differences in the kinetics of dissipation were observed for the herbicide a.i. sulcotrione alone and in the formulation Tangenta® [91]. For the determination of isoxaflutole at low concentrations, liquid chromatography coupled with tandem mass spectrometry (LC-MS/MS) is the analytical method of choice with LOD and LOQ values low enough to detect the a.i. under the maximum residue level in food and feed products or the maximum individual limit (0.1 μg/L) in surface water [92–95]. In our study, a simpler LC-UV method was applied for the detection of isoxaflutole dissipation, as our aim was only to determine its DT_{50} values, not the a.i. at very low concentrations. In an environmental and ecotoxicological impact assessment study of various herbicide mixtures of 20 a.i.s, carried out by an Italian researchers, isoxaflutole itself and combinations containing it were found to be most favorable regarding the pesticide index (a specific score based on selected pesticide indicators including water affinity, soil mobility, soil persistence, water persistence, percolation index, partition coefficient, and volatility), while among the worst ones regarding the priority index for surface and ground waters (a specific score based on sales data, type of application, environmental distribution, and soil persistence) [96]. A recent survey reported the use of isoxaflutole to be 193 tons/year in the US only in 2018, being the third-most highly used HPPD inhibitor in maize with a single application annually at an application dosage of 72 g/ha [29]. Lipophilicity of the two additives cyprosulfamide and 1,2-benzisothiazol-3(2H)-one are similar ($logK_{ow}$ 0.55 [44] and 0.70 [97], respectively); therefore, their potential effects on the solubility features of the two orders of magnitude and more lipophilic feature of isoxaflutole ($logK_{ow}$ 2.34 [18]) are similar, and the concentration of 1,2-benzisothiazol-3(2H)-one is 400–4000-fold lower than that of isoxaflutole.

5. Conclusions

The ecotoxicological effect of two isoxaflutole-based formulations (Merlin Flexx and Merlin WG75) and two additives (cyprosulfamide safener and 1,2-benzisothiazol-3(2H)-one antimicrobial agent) were determined on the growth inhibition of *R. subcapitata* green algal indicator species. The combined effects of the a.i. isoxaflutole with cyprosulfamide and cyprosulfamide + 1,2-benzisothiazol-3(2H)-one were also determined to obtain a better insight to the effect of herbicide formulations and their ingredients. After the ban of atrazine (in 2004 in the EU), isoxaflutole has become its replacement a.i.; thus, the application of isoxaflutole-based formulations is increasing. As a consequence, the environmental burden from the a.i. and additives applied in its formulations is on the rise as well. For risk assessments, it is essential to know the fate of the ingredients in different environmental matrices and the dissipation characteristics. Our results indicated that the additives in Merlin Flexx and Merlin WG75 influence the DT_{50} values in aqueous media and the toxicity of the isoxaflutole-based formulations. The toxicity of isoxaflutole was lower in the presence of the tested safener and antimicrobial agent presented in Merlin Flexx. Moreover, the presence of other additives in the formulation resulted in an even lower toxicity of the formulation. In contrast, the additives in the formulation prolonged the dissipation time of the a.i.; thus, the environmental load represented by the ingredients is longer if they are present together in the same media.

Results of ecotoxicity assays showed that the most sensitive endpoint was the induced fluorescent-based parameter Rfd* that characterizes the aptness of the photochemical system of the microalgae tested well. The benefit of the application of Rfd* values is that the determination requires no sample preparation compared to the measurement of Chl-a content. The least sensitive parameter was the optical density. However, the toxicity order of the tested compounds was the same for all the three endpoints. In this study, the corresponding OECD Guideline was applied for the detection of the ecotoxicological effects of additives in isoxaflutole-based herbicide formulations. Thus, the results are limited to the growth inhibition of test compounds on *R. subcapitata* determined by conventionally applied endpoints (OD and Chl-a content). In addition, a fluorescence-based parameter (Rfd*) was also assessed to obtain information regarding the possibly detrimental effects of the compounds and their combinations tested on the PSII photochemical system. For a more complex risk assessment study, investigation of the test compounds in field studies are required and biochemical and physiological experiments can contribute for a better understanding of the mode of action of test compounds.

Author Contributions: Conceptualization, L.K. and A.S. (András Székács); methodology, E.T., D.L., S.K., A.S. (Augustine Siakwa), L.K., A.B. and S.L.; software, A.B.; validation, D.L., E.T., S.K. and E.L.; investigation, E.T., D.L., A.S. (Augustine Siakwa), M.M. and S.K.; writing—original draft preparation, E.T., S.K. and M.M.; writing—review and editing, D.L., A.S. (Augustine Siakwa), L.K., A.B., E.L. and A.S. (András Székács); visualization, E.T.; supervision, A.S. (András Székács); funding acquisition, L.K. and A.S. (András Székács). All authors have read and agreed to the published version of the manuscript.

Funding: This work has been funded by the Hungarian National Research, Development and Innovation Office within the National Competitiveness and Excellence Program, project NVKP_16-1-2016-0049, "In situ, complex water quality monitoring by using direct or immunofluorimetry and plasma spectroscopy" (Aquafluosense 2017–2021), and project 2022-2.1.1-NL-2022-00006, "Development of the Agrotechnology National Laboratory" (Grant agreement NKFIH-3524-1/2022), supported by the National Research, Development and Innovation Fund by the Hungarian Ministry of Culture and Innovation, as well as project TKP2021-NVA-22 within the framework of the Thematic Excellence Program 2021, National Defense, National Security Sub-Program, and the Research Excellence Program 2024 (Kutatási Kiválósági Program 2024) of the Hungarian University of Agriculture and Life Sciences.

Institutional Review Board Statement: Not applicable.

Informed Consent Statement: Not applicable.

Data Availability Statement: The data presented in this study are available on request from the corresponding author. The data are not publicly available due to privacy reasons.

Acknowledgments: The authors express their sincere appreciation to Christoph Then, executive director of Testbiotech e.V. (https://www.testbiotech.org) for his whistleblowing precautionary warning regarding isoxaflutole during European Commission public debates regarding GMO regulation and at the conferences of the European Network of Scientists for Social and Environmental Responsibility (https://ensser.org) that directed the interest of our research group to the environmental features of this herbicide a.i. Thanks are also due to Dóra Gabnai for her technical assistance in the ecotoxicity tests, to Attila Csákányi for the mechanical design, and to László Lugosi for the motor control software.

Conflicts of Interest: The authors declare no conflict of interest.

References

1. Brock, T.C.M.; Arts, G.H.P.; Maltby, L.; Van den Brink, P.J. Aquatic risks of pesticides, ecological protection goals, and common aims in European Union legislation. *Integr. Environ. Assess. Manag.* **2006**, *2*, e20–e46. [CrossRef]
2. Arias-Estévez, M.; López-Periago, E.; Martínez-Carballo, E.; Simal-Gándara, S.; Mejuto, J.-C.; García-Río, L. The mobility and degradation of pesticides in soils and the pollution of groundwater resources. *Agric. Ecosyst. Environ.* **2008**, *123*, 247–260. [CrossRef]
3. Verro, R.; Finizio, A.; Otto, S.; Vighi, M. Predicting pesticide environmental risk in intensive agricultural areas. II: Screening level risk assessment of complex mixtures in surface waters. *Environ. Sci. Technol.* **2009**, *43*, 530–537. [CrossRef] [PubMed]
4. Nicolopoulou-Stamati, P.; Maipas, S.; Kotampasi, C.; Stamatis, P.; Hens, L. Chemical pesticides and human health: The urgent need for a new concept in agriculture. *Front. Public Health* **2016**, *4*, 148. [CrossRef] [PubMed]
5. Tang, F.H.M.; Lenzen, M.; McBratney, A.; Maggi, F. Risk of pesticide pollution at the global scale. *Nat. Geosci.* **2021**, *14*, 206–210. [CrossRef]
6. Tilman, D.; Fargione, J.; Wolff, B.; D'Antonio, C.; Dobson, A.; Howarth, R.; Schindler, D.; Schlesinger, W.H.; Simberloff, D.; Swackhamer, D. Forecasting agriculturally driven global environmental change. *Science* **2001**, *292*, 281–284. [CrossRef]
7. Tscharntke, T.; Clough, Y.; Wanger, T.C.; Jackson, L.; Motzke, I.; Perfecto, I.; Vandermeer, J.; Whitbread, A. Global food security, biodiversity conservation and the future of agricultural intensification. *Biol. Conserv.* **2012**, *151*, 53–59. [CrossRef]
8. Otero, I.; Farrell, K.N.; Pueyo, S.; Kallis, G.; Kehoe, L.; Haberl, H.; Plutzar, C.; Hobson, P.; García-Márquez, J.; Rodríguez-Labajos, B.; et al. Biodiversity policy beyond economic growth. *Conserv. Lett.* **2020**, *13*, e12713. [CrossRef]
9. Estrada-Carmona, N.; Sánchez, A.C.; Remans, R.; Jones, S.K. Complex agricultural landscapes host more biodiversity than simple ones: A global meta-analysis. *Proc. Natl. Acad. Sci. USA* **2022**, *119*, e2203385119. [CrossRef]
10. Hüesker, F.; Lepenies, R. Why does pesticide pollution in water persist? *Environ. Sci. Policy* **2022**, *128*, 185–193. [CrossRef]
11. Stehle, S.; Schulz, R. Agricultural insecticides threaten surface waters at the global scale. *Proc. Natl. Acad. Sci. USA* **2015**, *112*, 5750–5755. [CrossRef]

12. Organisation for Economic Co-Operation and Development. *OECD Compendium of Agri-Environmental Indicators*; OECD Publishing: Paris, France, 2013.
13. Pacanoski, Z. *Herbicides and adjuvants. In. Herbicides—Physiology of Action, and Safety*; Price, A., Kelton, J., Sarunaite, L., Eds.; IntechOpen: London, UK, 2015; pp. 125–147. [CrossRef]
14. Duhoux, A.; Pernin, F.; Desserre, D.; Délye, C. Herbicide safeners decrease sensitivity to herbicides inhibiting acetolactate-synthase and likely activate non-target-site-based resistance pathways in the major grass weed *Lolium* sp. (rye-grass). *Front. Plant Sci.* **2017**, *8*, 1310. [CrossRef]
15. Varga, Z.; Nicol, E.; Bouchonnet, S. Photodegradation of benzisothiazolinone: Identification and biological activity of degradation products. *Chemosphere* **2020**, *240*, 124862. [CrossRef]
16. U.S. Office of the Federal Register. *Code of Federal Regulations*; Title 40. Protection of Environment. Chapter I. Environmental Protection Agency. Subchapter E. Pesticide Programs. Part 180. Tolerances and Exemptions for Pesticide Chemical Residues in Food. Subpart D. Exemptions from Tolerances § 180.930. Inert Ingredients Applied To Animals; Exemptions from the Requirement of a Tolerance; U.S. Office of the Federal Register: Washington, DC, USA, 2024. Available online: https://www.ecfr.gov/current/title-40/chapter-I/subchapter-E/part-180/subpart-D/section-180.930 (accessed on 18 March 2024).
17. Klátyik, S.; Bohus, P.; Darvas, B.; Székács, A. Authorization and toxicity of veterinary drugs and plant protection products: Residues of the active ingredients in food and feed and toxicity problems related to adjuvants. *Front. Vet. Sci.* **2017**, *4*, 146. [CrossRef]
18. Turner, J.A. (Ed.) *The Pesticide Manual*, 19th ed.; British Crop Protection Council: Cambridge, UK, 2022; pp. 700–701.
19. Pallett, K.E.; Little, J.P.; Sheekey, M.; Veerasekaran, P. The mode of action of isoxaflutole: I. physiological effects, metabolism, and selectivity. *Pestic. Biochem. Physiol.* **1998**, *62*, 113–124. [CrossRef]
20. Pallett, K.E.; Cramp, S.M.; Little, J.P.; Veerasekaran, P.; Crudace, A.J.; Slater, A.E. Isoxaflutole: The background to its discovery and the basis of its herbicidal properties. *Pest Manag. Sci.* **2001**, *57*, 133–142. [CrossRef] [PubMed]
21. US Environmental Protection Agency. *Pesticide Fact Sheet. Isoxaflutole*; United States Environmental Protection Agency: Washington, DC, USA, 1998; pp. 1–15. Available online: https://www3.epa.gov/pesticides/chem_search/reg_actions/registration/fs_PC-123000_15-Sep-98.pdf (accessed on 18 March 2024).
22. Yang, D.; Wang, Y.-E.; Chen, M.; Liu, H.; Huo, J.; Zhang, J. Discovery of bis-5-cyclopropylisoxazole-4-carboxamides as novel potential 4-hydroxyphenylpyruvate dioxygenase inhibitors. *J. Agric. Food Chem.* **2023**, *71*, 5136–5142. [CrossRef] [PubMed]
23. Stephenson, D.O.; Bond, J.A. Evaluation of thiencarbazone-methyl- and isoxaflutole-based herbicide programs in corn. *Weed Technol.* **2012**, *26*, 37–42. [CrossRef]
24. Benoit, L.; Soltani, N.; Hooker, D.C.; Robinson, D.E.; Sikkema, P.H. Efficacy of HPPD-inhibiting herbicides applied preemergence or postemergence for control of multiple herbicide resistant waterhemp [*Amaranthus tuberculatus* (Moq.) Sauer]. *Can. J. Plant Sci.* **2019**, *99*, 379–383. [CrossRef]
25. Mausbach, J.; Irmak, S.; Sarangi, D.; Lindquist, J.; Jhala, A.J. Control of acetolactate synthase inhibitor/glyphosate-resistant Palmer amaranth (*Amaranthus palmeri*) in isoxaflutole/glufosinate/glyphosate-resistant soybean. *Weed Technol.* **2021**, *35*, 779–785. [CrossRef]
26. European Commission. Commission Implementing Regulation (EU) 2019/717 of 8 May 2019 renewing the approval of the active substance isoxaflutole in accordance with Regulation (EC) No 1107/2009 of the European Parliament and of the Council concerning the placing of plant protection products on the market, and amending the Annex to Commission Implementing Regulation (EU) No 540/2011. *Off. J. Eur. Union* **2019**, L 122/44. Available online: https://eur-lex.europa.eu/legal-content/EN/TXT/PDF/?uri=CELEX:32019R0717 (accessed on 18 March 2024).
27. European Commission. Isoxaflutole. SANTE/11653/2017 Rev 2. 22 March 2019. Final Renewal Report for the Active Substance Isoxaflutole Finalised in the Standing Committee on Plants, Animals, Food and Feed at Its Meeting on 22 March 2019 in View of the Renewal of the Approval of Isoxaflutole as an Active Substance in Accordance with Regulation (EC) No 1107/2009. Available online: https://eur-lex.europa.eu/legal-content/EN/TXT/PDF/?uri=CELEX:32019R0717&from=EN (accessed on 23 March 2024).
28. Foster, D.C.; Dotray, P.A.; Thompson, C.N.; Baldwin, G.B.; Moore, F.T. HPPD-resistant cotton response and weed management systems using isoxaflutole. *Weed Technol.* **2022**, *36*, 671–677. [CrossRef]
29. Jhala, A.J.; Kumar, V.; Yadav, R.; Jha, P.; Jugulam, M.; Williams, M.M., II; Hausman, N.E.; Dayan, F.E.; Burton, P.M.; Dale, R.P.; et al. 4-Hydroxyphenylpyruvate dioxygenase (HPPD)-inhibiting herbicides: Past, present, and future. *Weed Technol.* **2022**, *37*, 1–14. [CrossRef]
30. US Environmental Protection Agency. *Final Registration of Isoxaflutole on Isoxaflutole-Resistant Soybeans*; United States Environmental Protection Agency: Washington, DC, USA, 2020; 17p. Available online: https://www.regulations.gov/document/EPA-HQ-OPP-2019-0398-0071 (accessed on 18 March 2024).
31. Székács, A.; Darvas, B. Forty years with glyphosate. In *Herbicides—Properties, Synthesis and Control of Weeds*; Hasaneenm, M.N.A.E.-G., Ed.; InTech: Rijeka, Croatia, 2012; pp. 247–284. [CrossRef]
32. Székács, A.; Darvas, B. Re-registration challenges of glyphosate in the European Union. *Front. Environ. Sci.* **2018**, *6*, 78. [CrossRef]
33. Meyer, M.T.; Scribner, E.A.; Kalkhoff, S.J. Comparison of fate and transport of isoxaflutole to atrazine and metolachlor in 10 Iowa rivers. *Environ. Sci. Technol.* **2007**, *41*, 6933–6939. [CrossRef]

34. Rogers, J.; Chen, M.; Yang, K.; Graham, J.; Parker, K.M. Production of dichloroacetonitrile from derivatives of isoxaflutole herbicide during water treatment. *Environ. Sci. Technol.* **2023**, *57*, 18443–18451. [CrossRef]
35. Lázár, D.; Takács, E.; Mörtl, M.; Klátyik, S.; Barócsi, A.; Kocsányi, L.; Lenk, S.; Domján, L.; Szarvas, G.; Lengyel, E.; et al. Application of a fluorescence-based instrument prototype for chlorophyll measurements and its utility in an herbicide algal ecotoxicity assay. *Water* **2023**, *15*, 1866. [CrossRef]
36. Ramanarayanan, T.; Narasimhan, B.; Srinivasan, R. Characterization of fate and transport of isoxaflutole, a soil-applied corn herbicide, in surface water using a watershed model. *J. Agric. Food Chem.* **2005**, *53*, 8848–8858. [CrossRef]
37. Santos, E.A.; Correia, N.M.; Silva, J.R.M.; Velini, E.D.; Passos, A.B.R.J.; Durigan, J.C. Herbicide detection in groundwater in Córrego Rico-SP watershed. *Planta Daninha* **2015**, *33*, 147–155. [CrossRef]
38. Milan, M.; Ferrero, A.; Letey, M.; De Palo, F.; Vidotto, F. Effect of buffer strips and soil texture on runoff losses of flufenacet and isoxaflutole from maize fields. *J. Environ. Sci. Health B* **2013**, *48*, 1021–1033. [CrossRef]
39. Karim, R.; Reading, L.; Dawes, L.; Dahan, O.; Orr, G. Pesticide transport through the vadose zone under sugarcane in the Wet Tropics, Australia. *Soil* **2023**, *9*, 381–398. [CrossRef]
40. Franco, D.S.P.; Georgin, J.; Lima, E.C.; Silva, L.F.O. Advances made in removing paraquat herbicide by adsorption technology: A review. *J. Water Process Eng.* **2022**, *49*, 102988. [CrossRef]
41. Ogunbiyi, O.D.; Akamo, D.O.; Oluwasanmi, E.E.; Adebanjo, J.; Isafiade, B.A.; Ogunbiyi, T.J.; Alli, Y.A.; Ayodele, D.T.; Oladoye, P.O. Glyphosate-based herbicide: Impacts, detection, and removal strategies in environmental samples. *Groundw. Sust. Dev.* **2023**, *22*, 100961. [CrossRef]
42. Lozano, V.L.; Pizarro, H.N. Glyphosate lessons: Is biodegradation of pesticides a harmless process for biodiversity? *Environ. Sci. Eur.* **2024**, *36*, 55. [CrossRef]
43. McFadden, M.E.; Hladik, M.L. Cyprosulfamide: Analysis of the herbicide safener and two of its degradates in surface water and groundwater from the Midwestern United States. *ACS Agric. Sci. Technol.* **2021**, *1*, 355–361. [CrossRef]
44. Femi-Oloye, O.P.; Oloye, F.F.; Jones, P.D.; Giesy, J.P. Sorption behaviour and toxicity of an herbicide safener "cyprosulfamide". *Sci. Total Environ.* **2023**, *859*, 160077. [CrossRef]
45. Khan, S.J.; Murchland, D.; Rhodes, M.; Waite, T.D. Management of concentrated waste streams from high-pressure membrane water treatment systems. *Crit. Rev. Environ. Sci. Technol.* **2009**, *39*, 367–415. [CrossRef]
46. Bollmann, U.E.; Vollertsen, J.; Carmeliet, J.; Bester, K. Dynamics of biocide emissions from buildings in a suburban stormwater catchment—Concentrations, mass loads and emission processes. *Water Res.* **2014**, *56*, 66–76. [CrossRef] [PubMed]
47. Paun, I.; Pirvu, F.; Iancu, V.I.; Chiriac, F.L. Occurrence and transport of isothiazolinone-type biocides from commercial products to aquatic environment and environmental risk assessment. *Int. J. Environ. Res. Public Health* **2022**, *19*, 7777. [CrossRef] [PubMed]
48. Butterwick, C.; Heaney, S.I.; Talling, J.F. A comparison of eight methods for estimating the biomass and growth of planktonic algae. *Br. Phycol. J.* **1982**, *17*, 69–79. [CrossRef]
49. Berden-Zrimec, M.; Drinovec, L.; Zrime, A. Delayed Fluorescence. In *Chlorophyll a Fluorescence in Aquatic Sciences: Methods and Applications*, 1st ed.; Suggett, D.J., Ondrej, P., Borowitzka, M.A., Eds.; Springer: Dordrecht, The Netherlands, 2010; Volume 4, pp. 293–309. [CrossRef]
50. Lenk, S.; Gádoros, P.; Kocsányi, L.; Barócsi, A. Teaching laser-induced fluorescence of plant leaves. *Eur. J. Phys.* **2016**, *37*, 064003. [CrossRef]
51. Z8 Medium. Available online: https://www-cyanosite.bio.purdue.edu/media/table/Z8.html (accessed on 18 March 2024).
52. Merlin Flexx; Number of MSDS: 102000016788; Bayer Hungária Kft. Bayer Crop Science: Budapest, Hungary, 2022. Available online: https://ipad.bayerhungaria.hu/cropscience/content/gazdainfo/v2/online/pdf_79_termekek_biztonsagtechnikaiokirat/pdf_79_termekek_biztonsagtechnikaiokirat.pdf?986709 (accessed on 18 March 2024).
53. Merlin WG75; Number of MSDS: 102000001698; Bayer (Pty) Ltd.: Isando, South Africa. 2019. Available online: https://www.cropscience.bayer.africa/content/dam/bayer-crop-science/south-africa/bcs/product-msds/merlin_msds.pdf (accessed on 18 March 2024).
54. European Parliament and the Council. Regulation (EC) No 1907/2006 of the European Parliament and of the Council of 18 December 2006 concerning the Registration, Evaluation, Authorisation and Restriction of Chemicals (REACH), establishing a European Chemicals Agency. *Off. J. Eur. Union* **2006**, 278. Available online: https://eur-lex.europa.eu/LexUriServ/LexUriServ.do?uri=OJ:L:2007:136:0003:0280:en:PDF (accessed on 18 March 2024).
55. Organisation for Economic Cooperation and Development. *Freshwater Alga and Cyanobacteria, Growth Inhibition Test (201)*; OECD Guidelines for the Testing of Chemicals, Section 2; OECD: Paris, France, 2011; Volume 2.
56. International Organization for Standardization. 10260:1992 Water Quality—Measurement of Biochemical Parameters—Spectrometric Determination of the Chlorophyll-a Concentration. International Organization for Standardization: Geneva, Switzerland. Available online: https://www.iso.org/standard/18300.html (accessed on 18 March 2024).
57. Felföldy, L. A biológiai vízminősítés. 4. javított és bővített kiadás. In *Vízügyi Hidrobiológia*; VGI: Budapest, Hungary, 1987; Volume 16, 258p.
58. Barócsi, A.; Kocsányi, L.; Várkonyi, S.; Richter, P.; Csintalan, Z.; Szente, K. Two-wavelength, multipurpose, truly portable chlorophyll fluorometer and its application in field monitoring of phytoremediation. *Meas. Sci. Technol.* **2000**, *11*, 717–729. [CrossRef]

59. Barócsi, A.; Lenk, S.; Kocsányi, L.; Buschmann, C. Excitation kinetics during induction of chlorophyll a fluorescence. *Photosynthetica* **2009**, *47*, 104–111. [CrossRef]
60. Kautsky, H.; Hirsch, A. Neue Versuche zur Kohlensäureassimilation. *Naturwissenschaften* **1931**, *19*, 964. [CrossRef]
61. Huang, Y.; Thomson, S.J.; Molin, W.T.; Reddy, K.N.; Yao, H. Early detection of soybean plant injury from glyphosate by measuring chlorophyll reflectance and fluorescence. *J. Agric. Sci.* **2012**, *4*, 117–124. [CrossRef]
62. Vredenberg, W.J. On the quantitative relation between dark kinetics of NPQ-induced changes in variable fluorescence and the activation state of the CF0·CF1·ATPase in leaves. *Photosynthetica* **2018**, *56*, 139–149. [CrossRef]
63. Da Silva Santarossa, M.A.; Coleone, A.C.; de Mello, N.P.; Ignácio, N.F.; Machado, A.A.; Marques Silva, J.R.; Velini, E.D.; Machado Neto, J.G. Contamination of fee-fishing ponds with agrochemicals used in sugarcane crops. *SN Appl. Sci.* **2020**, *2*, 1498. [CrossRef]
64. Wisconsin Department of Agriculture, Trade and Consumer Protection Agricultural Resource Management Division. Final Environmental Impact Statement for the Use of Pesticides Containing Isoxaflutole in Wisconsin. 2002. Available online: https://www.fluoridealert.org/wp-content/pesticides/isoxaflutole.wisc.feis.2002.pdf (accessed on 18 March 2024).
65. European Food Safety Authority. Peer review of the pesticide risk assessment of the active substance isoxaflutole. *EFSA J.* **2016**, *14*, 4416. [CrossRef]
66. Pesticide Properties DataBase. Isoxaflutole. Available online: http://sitem.herts.ac.uk/aeru/ppdb/en/Reports/412.htm (accessed on 18 March 2024).
67. Jeong, S.-M.; Noh, T.-K.; Kim, D.-S. Herbicide bioassay using a multi-well plate and plant spectral image analysis. *Sensors* **2024**, *24*, 919. [CrossRef]
68. National Registration Authority for Agricultural and Veterinary Chemicals (Australia). *Public Release Summary on Evaluation of the New Active Isoxaflutole in the Product Balance 750WG Herbicide*; The Authority: Canberra, Australia, 2001; 35p.
69. Nagy, K.; Duca, R.C.; Lovas, S.; Creta, M.; Scheepers, P.; Godderis, L.; Ádám, B. Systematic review of comparative studies assessing the toxicity of pesticide active ingredients and their product formulations. *Environ. Res.* **2020**, *181*, 108926. [CrossRef]
70. Mesnage, R.; Defarge, N.; Spiroux de Vendômois, J.; Séralini, G.E. Major pesticides are more toxic to human cells than their declared active principles. *Biomed. Res. Int.* **2014**, *2014*, 179691. [CrossRef] [PubMed]
71. Bringolf, R.B.; Cope, W.G.; Barnhart, M.C.; Mosher, S.; Lazaro, P.R.; Shea, D. Acute and chronic toxicity of pesticide formulations (atrazine, chlorpyrifos, and permethrin) to glochidia and juveniles of *Lampsilis siliquoidea*. *Environ. Toxicol. Chem.* **2007**, *26*, 2101–2107. [CrossRef]
72. Takács, E.; Klátyik, S.; Mörtl, M.; Rácz, G.; Kovács, K.; Darvas, B.; Székács, A. Effects of neonicotinoid insecticide formulations and their components on *Daphnia magna*—The role of active ingredients and co-formulants. *Int. J. Environ. Anal. Chem.* **2017**, *97*, 885–900. [CrossRef]
73. Seralini, G.-E. Pesticides in formulations: New Revolutionary Findings. *Toxics* **2024**, *12*, 151. [CrossRef] [PubMed]
74. Defarge, N.; Takács, E.; Lozano, V.L.; Mesnage, R.; Spiroux de Vendômois, J.; Séralini, G.E.; Székács, A. Co-Formulants in Glyphosate-Based Herbicides Disrupt Aromatase Activity in Human Cells below Toxic Levels. *Int. J. Environ. Res. Public Health* **2016**, *13*, 264. [CrossRef] [PubMed]
75. Song, H.Y.; Kim, Y.H.; Seok, S.J.; Gil, H.W.; Yang, J.O.; Lee, E.Y.; Hong, S.Y. Cellular toxicity of surfactants used as herbicide additives. *J. Korean Med. Sci.* **2012**, *27*, 3–9. [CrossRef] [PubMed]
76. Wang, W.; Zhang, J.; Wu, J.; Yu, R.; Zhang, Y.; Sun, L.; Gao, Y. Acute Toxicity and Ecotoxicological Risk Assessment of Three Volatile Pesticide Additives on the Earthworm—Eisenia fetida. *Int. J. Environ. Res. Public Health* **2021**, *18*, 11232. [CrossRef]
77. Klátyik, S.; Takács, E.; Mörtl, M.; Földi, A.; Trábert, Z.; Ács, É. Dissipation of the herbicide active ingredient glyphosate in natural water samples in the presence of biofilms. *Int. J. Environ. Anal. Chem.* **2017**, *97*, 901–921. [CrossRef]
78. Oláh, M.; Farkas, E.; Székács, I.; Horváth, R.; Székács, A. Cytotoxic effects of Roundup Classic and its components on NE-4C and MC3T3-E1 cell lines determined by biochemical and flow cytometric assays. *Toxicol. Rep.* **2022**, *9*, 914–926. [CrossRef]
79. Székács, I.; Farkas, E.; Gémes, B.; Takács, E.; Székács, A.; Horváth, R. Integrin targeting of glyphosate and its cell adhesion modulation effects on osteoblastic MC3T3-E1 cells revealed by label-free optical biosensing. *Sci. Rep.* **2018**, *8*, 17401. [CrossRef]
80. European Commission. Commission Implementing Regulation (EU) 2016/1313 of 1 August 2016 amending Implementation Regulation (EU) No 540/2011 as regards the conditions of approval of the active substance glyphosate. *Off. J. Eur. Union* **2016**, L 208/1. Available online: https://eur-lex.europa.eu/legal-content/EN/TXT/HTML/?uri=CELEX:32016R1313 (accessed on 18 March 2024).
81. Mesnage, R.; Benbrook, C.; Antoniou, M.N. Insight into the confusion over surfactant co-formulants in glyphosate-based herbicides. *Food Chem. Toxicol.* **2019**, *128*, 137–145. [CrossRef] [PubMed]
82. US Environmental Protection Agency. *Memorandum. Carcinogenicity peer Review of Isoxaflutole. Office of Prevention, Pesticides, and Toxic Substances*; US Environmental Protection Agency: Washington, DC, USA, 1997. Available online: https://www3.epa.gov/pesticides/chem_search/cleared_reviews/csr_PC-123000_6-Aug-97_075.pdf (accessed on 18 March 2024).
83. US Environmental Protection Agency. 0 CFR Part 180 [EPA–HQ–OPP–2010–0845; FRL–8885–8] Isoxaflutole; Pesticide Tolerances. *Fed. Regist.* **2011**, *76*, 76309–76314. Available online: https://www.govinfo.gov/content/pkg/FR-2011-12-07/pdf/2011-31397.pdf (accessed on 18 March 2024).
84. Marx-Stoelting, P.; Niemann, L.; Ritz, V.; Ulbrich, B.; Gall, A.; Hirsch-Ernst, K.I.; Pfeil, R.; Solecki, R. Assessment of three approaches for regulatory decision making on pesticides with endocrine disrupting properties. *Regul. Toxicol. Pharmacol.* **2014**, *70*, 590–604. [CrossRef]

85. Mesnage, R.; Antoniou, M. Mammalian toxicity of herbicides used in intensive GM crop farming. In *Herbicides*; Mesnage, R., Zaller, J., Eds.; Elsevier: Amsterdam, The Netherlands, 2021; pp. 143–180. [CrossRef]
86. Wang, X.X.; Zhang, T.Y.; Dao, G.H.; Hu, H.Y. Interaction between 1,2-benzisothiazol-3(2H)-one and microalgae: Growth inhibition and detoxification mechanism. *Aquat. Toxicol.* **2018**, *205*, 66–75. [CrossRef] [PubMed]
87. Chamsi, O.; Pinelli, E.; Faucon, B.; Perrault, A.; Lacroix, L.; Sanchez-Perez, J.-M.; Charcosset, J.-Y. Effects of herbicide mixtures on freshwater microalgae with the potential effect of a safener. *Ann. Limnol. Int. J. Lim.* **2019**, *55*, 3. [CrossRef]
88. Mörtl, M.; Takács, E.; Klátyik, S.; Székács, A. Aquatic toxicity and loss of linear alkylbenzenesulfonates alone and in a neonicotinoid insecticide formulation in surface water. *Sci. Total Environ.* **2019**, *652*, 780–787. [CrossRef]
89. Du-Carrée, J.L.; Cabon, J.; Morin, T.; Danion, M. Immunological and metabolic effects of acute sublethal exposure to glyphosate or glyphosate-based herbicides on juvenile rainbow trout, *Oncorhynchus mykiss*. *Sci. Total Environ.* **2021**, *784*, 147162. [CrossRef]
90. Birolli, W.G.; Porto, A.L.M. Esfenvalerate biodegradation by marine fungi is affected by seawater and emulsifier formulation. *Environ. Sci. Poll. Res.* **2023**, *30*, 38394–38408. [CrossRef]
91. Šojić, D.V.; Orčić, D.Z.; Četojević-Simin, D.D.; Banić, N.D.; Abramović, B.F. Efficient removal of sulcotrione and its formulated compound Tangenta® in aqueous TiO_2 suspension: Stability, photoproducts assessment and toxicity. *Chemosphere* **2015**, *138*, 988–994. [CrossRef]
92. European Parliament and Council. Directive 2008/105/EC of the European Parliament and of the Council of 16 December 2008 on environmental quality standards in the field of water policy, amending and subsequently repealing Council Directives 82/176/EEC, 83/513/EEC, 84/156/EEC, 84/491/EEC, 86/280/EEC and amending Directive 2000/60/EC of the European Parliament and of the Council. *Off. J. Eur. Union* **2008**, *51*, 84–97. Available online: https://eur-lex.europa.eu/legal-content/EN/TXT/PDF/?uri=CELEX:32008L0105 (accessed on 23 March 2024).
93. Lan, F.; Li, Q.; Sun, L.; Yu, W.; Jiang, W.; Wang, Z.; Liu, C. Simultaneous determination of isoxaflutole and its two metabolites in corn under field conditions by LC–MS/MS. *J. Sci. Food Agric.* **2022**, *102*, 3480–3486. [CrossRef] [PubMed]
94. Goel, V.; Pandey, D.; Shukla, S. Multiresidue analysis and probabilistic dietary risk assessment of 241 pesticides in wheatgrass (*Triticum* sp.) using LC–MS/MS in combination with QuEChERS extraction. *Biomed. Chromatogr.* **2022**, *36*, e5411. [CrossRef] [PubMed]
95. Cao, J.; Pei, T.; Wang, Y.; Qin, S.; Qi, Y.; Ren, P.; Li, J. Terminal residue and dietary risk assessment of atrazine and isoxaflutole in corn using high-performance liquid chromatography–tandem mass spectrometry. *Molecules* **2023**, *28*, 7225. [CrossRef]
96. Vidotto, F.; Fogliatto, S.; Milan, M. A new and integrated approach to evaluate the environmental and ecotoxicological impact of herbicide mixtures: A case study in maize. *Sci. Total Environ.* **2022**, *842*, 156862. [CrossRef]
97. European Chemicals Agency. *1,2-Benzisothiazol-3(2H)-One. Scientific Properties*; European Chemicals Agency: Helsinki, Finland, 2023. Available online: https://echa.europa.eu/hu/brief-profile/-/briefprofile/100.018.292 (accessed on 18 March 2024).

toxics

Article

Preparation and Comprehensive Evaluation of the Efficacy and Safety of Chlorantraniliprole Nanosuspension

Xiquan Ding [1], Liang Guo [2], Qian Du [1], Tingyu Wang [1], Zhanghua Zeng [1], Yan Wang [1], Haixin Cui [1], Fei Gao [1,*] and Bo Cui [1,*]

[1] Institute of Environment and Sustainable Development in Agriculture, Chinese Academy of Agricultural Sciences, Beijing 100081, China
[2] College of Pharmacy, Henan University of Chinese Medicine, Zhengzhou 450046, China
* Correspondence: gaofei@caas.cn (F.G.); cuibo@caas.cn (B.C.)

Abstract: Chlorantraniliprole is a broad-spectrum insecticide that has been widely used to control pests in rice fields. Limited by its low solubility in both water and organic solvents, the development of highly efficient and environmentally friendly chlorantraniliprole formulations remains challenging. In this study, a low-cost and scalable wet media milling technique was successfully employed to prepare a chlorantraniliprole nanosuspension. The average particle size of the extremely stable nanosuspension was 56 nm. Compared to a commercial suspension concentrate (SC), the nanosuspension exhibited superior dispersibility, as well as superior foliar wetting and retention performances, which further enhanced its bioavailability against *Cnaphalocrocis medinalis*. The nanosuspension dosage could be reduced by about 40% while maintaining a comparable efficacy to that of the SC. In addition, the chlorantraniliprole nanosuspension showed lower residual properties, a lower toxicity to non-target zebrafish, and a smaller effect on rice quality, which is conducive to improving food safety and the ecological safety of pesticide formulations. In this work, a novel pesticide-reduction strategy is proposed, and theoretical and data-based support is provided for the efficient and safe application of nanopesticides.

Keywords: chlorantraniliprole; nanosuspension; efficacy; safety

1. Introduction

Chlorantraniliprole is an anthranilic diamide insecticide (Figure 1) that has been widely used in rice fields to control *Cnaphalocrocis medinalis, Nilaparvata lugens Stal, Chilo suppressalis*, and other pests. It has both contact and stomach poison activities [1,2]. The insecticidal mechanism of chlorantraniliprole involves the activation of the ryanodine receptor, leading to the discharge of intracellular calcium in insects' smooth and striated muscle cells, which causes muscular regulation weakness and paralysis, culminating in pests' deaths [3,4]. However, the low solubility of chlorantraniliprole in both water (1.023 mg/L) and organic solvents (3.446 mg/L in acetone, 1.714 mg/L in methanol, 0.711 mg/L in acetonitrile, 1.144 mg/L in ethyl acetate at 20–25 °C) has greatly hindered the development of highly efficient and environmentally friendly formulations. Nowadays, chlorantraniliprole is mainly formulated as a suspension concentrate (SC) and water-dispersible granule (WDG). For conventional formulations, dust drift and roll down result in the loss of >50% of the applied pesticides during field application [5,6]. In addition, organic solvents and excessive amounts of emulsifiers are used to maintain effective formulation performance. The low efficacy and the composition of conventional formulations lead to resource waste, pest resistance, and a series of food safety and environmental pollution issues [7–9]. Therefore, there is an urgent need to develop a green and scalable strategy to build an environmentally friendly chlorantraniliprole delivery system that is highly efficacious and capable of tackling the problems related to the use of sparingly soluble pesticides.

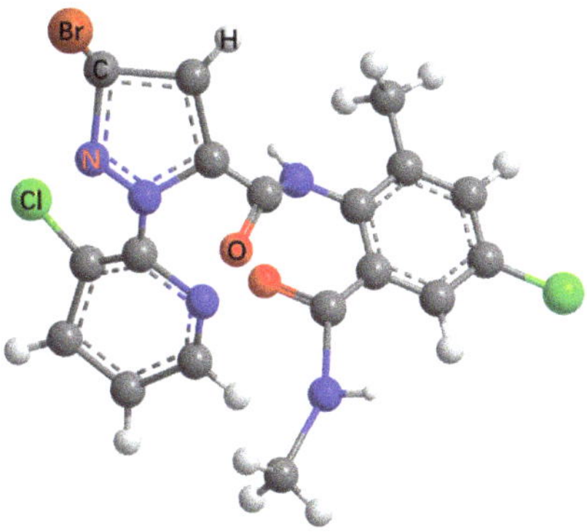

Figure 1. Stereochemical structure of chlorantraniliprole.

According to the Noyes–Whitney equation, a reduction in particle size can increase the saturation solubility and dissolution rate of materials [9], thereby improving the dispersibility of poorly soluble pesticide compounds. It has been widely demonstrated that nanotechnology can enhance the solubility, dispersibility, targeting, bioavailability, and safety of sparingly soluble pesticides, which has caused them to attract significant interest in the design and construction of nanopesticides [10–13]. Currently, there are two methods for preparing nanopesticides, namely, bottom-up and top-down approaches. Bottom-up approaches, such as micro-precipitation and solvent evaporation, involves the aggregation of molecules into nanoparticles [14,15]. Top-down approaches rely on mechanical forces to process larger particles into nanoscale particles. The typical top-down processes are media milling, high-pressure homogenization, dry co-grinding, and melt emulsification [16–23]. Patel et al. concluded that top-down techniques could significantly reduce drug particle sizes, increase their specific surface areas, and accelerate their dissolution rates, with their final formulations exhibiting good physical stability compared to conventional formulations [23]. Chin et al. produced a carbendazim nanosuspension via media milling which exhibited a superior stability and a 13.5% increase in insecticidal efficiency over a micron-sized suspension [24]. Francesco et al. prepared a zoxamid nanosuspension using the same method and increased the solubility 1.6-fold compared to the SC formulation while improving the deposition and retention abilities of the active ingredient on tomato plants [25]. In our previous work, we successfully constructed lambda–cyhalothrin and abamectin nanosuspensions using melt emulsification–high-pressure homogenization and wet milling methods, proving the effectiveness of nanotechnology for enhancing the dispersibility and biological activity of poorly soluble pesticides [26–28].

It is imperative to emphasize that recent studies on nanopesticides primarily concern formulation characterization and biological activity evaluations, and there is still a relative paucity of potential risk assessments on target and non-target organisms. In the mid-20th century, the widespread use of organochlorine pesticides such as DDT and lindane led to severe air, soil, and water pollution and caused the death of plants and animals through accumulation effects. Studies have linked the increasing incidence of nervous system damage like Alzheimer's disease, cardiovascular diseases, and diabetes to the consumption of pesticide-contaminated food [29,30]. Some types of pesticides are carcinogenic and mutagenic, seriously affecting human genes and causing harm to the growth and development of future generations [31]. Nanopesticides are likely to settle in water and remain in sediments, causing toxicity to aquatic organisms, and to accumulate in fish through ingestion, thus affecting human health through the food chain [32,33]. In addition, pesticides may damage the nutritional value of crops, reducing the mineral content and inhibiting synthesis of proteins, sugars, and vitamins [34]. Rico et al. found that CeO_2 nanoparticles can reduce the content of two essential elements (Fe and S), proteins (prolamin and glutelin), fatty acids (lauric and valeric acids), and starch in three rice varieties (high-, medium-, and low-amylose) [35]. Zhao et al. found that thifluzamide can change the chlorophyll, phenol, flavonoid, and protein contents in rice plants and that thifluzamide-loaded mesoporous silica nanoparticles could relieve the damage caused by thifluzamide to rice seedlings [36]. However, there are still relatively few reports focusing on both the nanopesticide residues and their impacts on food quality.

It is crucial to systematically evaluate the effect of nanopesticides on non-target organisms and foodstuffs while improving the effectiveness of pesticides. The potential environmental risk is a key issue, constraining the development and application of novel pesticide nanoformulations as well as the improvement of pesticide regulatory regimes. In this study, a facile and low-cost media milling technique was employed to prepare a chlorantraniliprole nanosuspension. The nanosuspensions' physicochemical properties and field efficacy were systematically compared with a conventional SC to reveal the enhancing effect of the nanoformulation and its underlying mechanism. In particular, in this study, we also focused on nanopesticides' safety issues from multiple perspectives related to pesticide residues in rice leaves and grains, effects on rice quality, and toxicity to non-target zebrafish. We propose a pesticide-reduction strategy and provide a theoretical basis and technical support for the efficient and safe application of nanopesticides.

2. Results and Discussion

2.1. Optimization of the Preparation Parameters of the Chlorantraniliprole Nanosuspension

According to our previous research, a composite surfactant composed of MRES and polycarboxylate (1:1, w/w) can provide effective electrostatic repulsion and a steric stabilization effect to stabilize the chlorantraniliprole nanoparticles produced by high-pressure homogenization combined with lyophilization. Therefore, the same surfactant combination was also applied for the preparation of the chlorantraniliprole nanosuspension by wet media milling. The surfactant content affects the amount of adsorption on the surface of the poor-water-soluble pesticide and further influences the formulation's dispersibility and stability. As shown in Table 1, an increase in surfactant content can reduce the particle size to a certain extent, but this effect was no longer evident when the surfactant amount was 50%. Therefore, a 1:2 (w/w) surfactant/pesticide ratio was used to prepare the chlorantraniliprole nanosuspension. In addition, the milling parameters, especially the milling time, have a crucial impact on the particle size and distribution of the resultant suspensions. As shown in Table 1, the milling time significantly affected the particle sizes of the chlorantraniliprole nanosuspensions, although it had essentially no effect on the PDI value. As the milling time increased from 0.5 h to 1 h, the average particle size of the chlorantraniliprole nanosuspension decreased from 288.4 nm to 244.2 nm, and when the milling time increased to 2 h, the particle size continued to decrease to 161.9 nm but increased again to 181.0 nm as the time was extended to 3 h. The reason for this is that a short milling period does not provide sufficient energy to break up large particles, while excessive milling generates a large amount of heat and induces particle agglomeration. Hence, 2 h was chosen as the optimal milling time for synthesizing the chlorantraniliprole nanosuspension.

Table 1. Effects of surfactant content and milling time on the particle size and distribution of the chlorantraniliprole nanosuspensions.

		Size (nm)	PDI
Surfactant/pesticide ratio	30%	167.1 ± 1.2	0.174 ± 0.013
	50%	161.9 ± 0.2	0.186 ± 0.010
	70%	160.3 ± 0.8	0.153 ± 0.007
Milling time (h)	0.5	288.4 ± 2.9	0.253 ± 0.009
	1	244.2 ± 3.1	0.174 ± 0.024
	2	161.9 ± 0.2	0.186 ± 0.010
	3	181.0 ± 1.4	0.176 ± 0.015

2.2. Particle Size and Zeta Potential

Particle size, distribution, and zeta potential are crucial indicators for evaluating the stability of water-based formulations. PDI values lower than 0.3 suggest a narrow size distribution, and a smaller PDI signifies a more uniform system. PDI values higher than 0.3 indicate that the system is susceptible to destabilization phenomena like Ostwald ripening and sedimentation [37,38]. As depicted in Table 2, the average particle size and PDI of the chlorantraniliprole nanosuspension were 161.9 ± 0.2 nm and 0.186 ± 0.010, respectively, measured using dynamic light scattering. In contrast, the particle size and PDI of the conventional SC formulation were 678.7 ± 27.3 nm and 0.446 ± 0.021, respectively, 4.2 and 2.4 times higher than that of the nanosuspension. Following the Ostwald Freundlich and Noyes Whitney equations, as the particle size decreases, the saturation solubility and dissolution rate of substances increase [39–41]. Consequently, the nanosuspension is more conducive to dissolving and dispersing poorly soluble pesticides in water, thereby mitigating the concentration discrepancies and reducing the incidences of chemical damage during spraying.

Table 2. Particle size, distribution, and zeta potential values of the chlorantraniliprole nanosuspension and SC.

Formulation	Size (nm)	PDI	Zeta Potential (mV)
Nanosuspension	161.9 ± 0.2 b	0.186 ± 0.010 b	−43.1 ± 0.7 b
SC	678.7 ± 27.3 a	0.446 ± 0.021 a	−29.4 ± 0.3 a

Different letters (a,b) in the table indicate significant differences at $p < 0.05$.

The potential polarity hinges upon the nature of the charge at the particle interface and is closely linked to the type and quantity of surfactants. The zeta potential of the chlorantraniliprole nanosuspension was measured to be −43.1 ± 0.7 mV, while the value of SC was −29.4 ± 0.3 mV (Table 2). Typically, an absolute zeta potential greater than 30 mV is indicative of system stability [42]. In this investigation, the highly negative zeta potential of the nanosuspension demonstrates that the anionic surfactants were indeed adsorbed on the pesticide surface and provided strong electrostatic repulsion against aggregation, thus enhancing the system stability [43].

2.3. Morphology

The morphologies of the chlorantraniliprole TC, SC, and nanosuspension are presented in Figure 2A–C. Due to the poor solubility and dispersibility of chlorantraniliprole in water, the TC exhibited an irregular blocky structure with micron-sized dimensions. Although the SC particle size was lower than that of the TC, it was still unevenly distributed, with micron and nanoparticles co-existing and noticeable aggregation. The suspension is a thermodynamically unstable system, and its long-term stabilization is an uphill battle against thermodynamics [41]. According to the Gibbs–Thomson equation [6,41] and the principle of thermodynamic energy reduction, the chlorantraniliprole SC with a broad size distribution is prone to undergo Ostwald ripening, leading to clarification, sedimentation, and agglomeration [44]. In contrast, the chlorantraniliprole nanosuspension produced through wet media milling exhibited a uniform and regular block-like structure. Based on SEM images, we found that the statistical particle size ranged from 34 nm to 86 nm, with an average diameter of 56 nm (Figure 2D). This value was smaller than that measured by DLS. This difference is due to the fact that SEM reflects the monodispersed particle size in a dried state, whereas DLS reflects the hydrodynamic diameter with hydrated or diffuse layers on the particle periphery [45]. Additionally, larger particles or aggregates in the suspension may shield the signals of smaller particles, leading to an overestimation of measurement values [46–49].

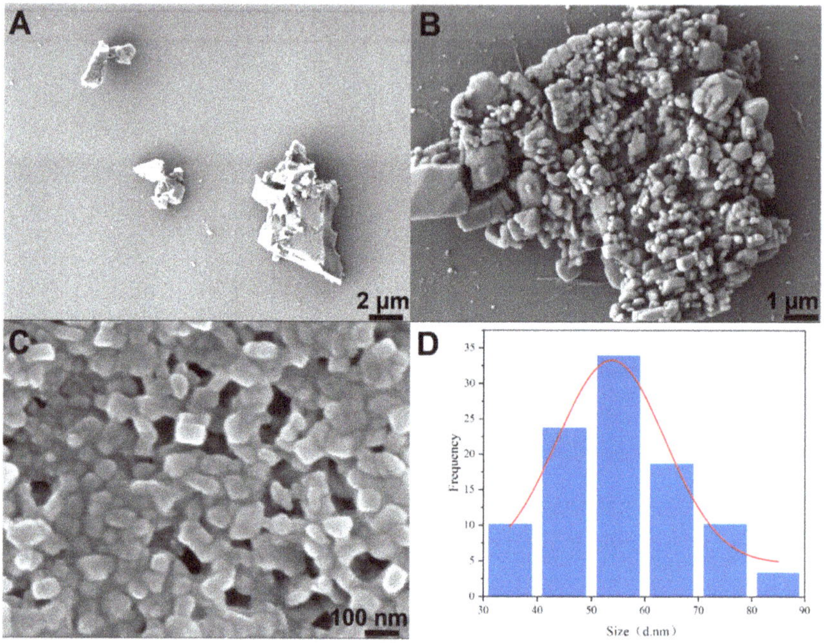

Figure 2. Morphologies of the chlorantraniliprole (**A**) TC, (**B**) SC, and (**C**) nanosuspension. (**D**) The statistical particle size of the nanosuspension (based on SEM images).

2.4. Stability

As shown in Figure 3, diluted solutions of the chlorantraniliprole nanosuspension at different concentrations exhibited a light blue color due to the Tyndall effect, whereas the SC was milky white at high concentrations. The storage stability of the nanosuspension was further investigated according to CIPAC MT 39, CIPAC MT 46, and GB/T 19136–2003. As shown in Table 3, the average particle size of the nanosuspension increased from 161.9 nm to 182.5 nm and 182.3 nm after 14 days of storage at 4 °C and 54 °C, respectively. However, the particle size changed only slightly when stored at 25 °C. Meanwhile, the PDI value remained below 0.2 during storage. This indicates that the nanosuspension had an excellent physical stability, which can be attributed to the strong electrostatic repulsion combined with the steric stabilization provided by the two polymer surfactants [50,51]. Therefore, reducing the particle size of chlorantraniliprole to the nanoscale is an effective way to improve its stability and shelf life.

Table 3. Storage stability of the chlorantraniliprole nanosuspension at 4 °C, 25 °C, and 54 °C.

Time (Day)	4 °C		25 °C		54 °C	
	Size (nm)	PDI	Size (nm)	PDI	Size (nm)	PDI
0	161.9 ± 0.2 e	0.186 ± 0.010 a	161.9 ± 0.2 bc	0.186 ± 0.010 a	161.9 ± 0.2 e	0.186 ± 0.010 a
2	166.1 ± 0.6 d	0.143 ± 0.006 e	159.0 ± 1.3 c	0.141 ± 0.008 b	156.7 ± 0.5 f	0.163 ± 0.004 ab
4	166.4 ± 0.6 d	0.137 ± 0.005 e	162.9 ± 1.5 bc	0.134 ± 0.008 b	173.9 ± 1.3 c	0.167 ± 0.002 ab
6	176.1 ± 0.4 c	0.159 ± 0.005 cd	164.6 ± 2.8 ab	0.147 ± 0.003 b	169.4 ± 0.3 d	0.149 ± 0.007 bc
8	175.8 ± 0.2 c	0.166 ± 0.011 bc	169.3 ± 3.2 a	0.146 ± 0.019 b	169.8 ± 0.7 d	0.133 ± 0.012 cd
10	177.5 ± 1.0 bc	0.148 ± 0.009 de	167.9 ± 2.6 a	0.151 ± 0.012 b	178.3 ± 2.5 b	0.165 ± 0.010 ab
12	178.8 ± 1.4 b	0.174 ± 0.007 ab	166.9 ± 2.6 ab	0.141 ± 0.008 b	176.4 ± 0.7 b	0.117 ± 0.023 d
14	182.5 ± 0.4 a	0.168 ± 0.007 bc	166.1 ± 0.8 ab	0.134 ± 0.017 b	182.3 ± 0.3 a	0.168 ± 0.012 ab

Different letters (a,b,c,d,e,f) in the table indicate significant differences at $p < 0.05$.

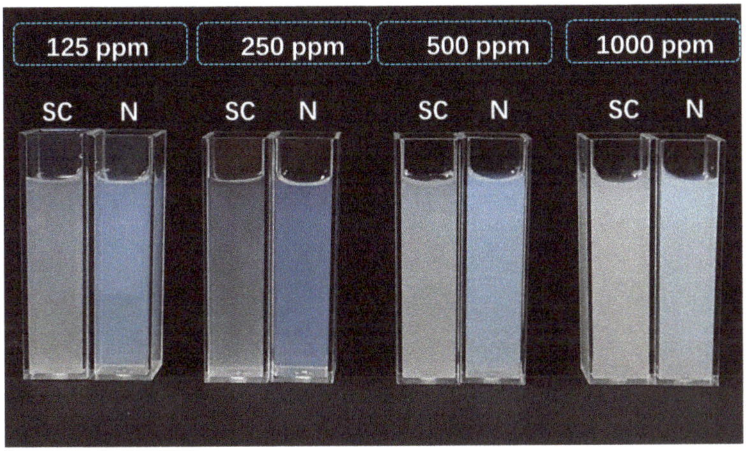

Figure 3. Appearance of the chlorantraniliprole nanosuspension(N) and SC diluted to different concentrations.

2.5. Foliar Wettability and Retention

The formulations' foliar wettabilities and retention influence the pesticide efficacy by affecting spreading, adhesion, and droplet persistence on crop leaves after spraying [52,53]. Under different application scenarios, the pesticide formulations will be diluted to different concentrations, so the foliar wettabilities of different diluents of the chlorantraniliprole nanosuspension on rice leaf surfaces were investigated. Hydrophobic leaves such as rice leaves are more difficult to wet and cannot retain water as well as hydrophilic leaves, so it is practical to enhance the pesticide utilization rate on hydrophobic surfaces. As shown in Figure 4A, the contact angle of the chlorantraniliprole SC on a rice leaf was 125.8°, which was comparable to the 18 g a.i./hm^2 dispersion of the nanosuspension. However, the contact angle of the nanosuspension dispersion continued to decrease with increasing pesticide concentrations, dropping to 120.7° at 24 g a.i./hm^2 and 116.9° at 30 g a.i./hm^2, significantly lower than the contact angle of the SC. A reduction in the contact angle is conducive to increasing the foliar wettability of pesticide droplets, which further enhances retention performance. As shown in Figure 4B, the retention of the chlorantraniliprole nanosuspension on rice leaves was 55.9% higher than that of the SC. In order to explore the reasons for the differences in deposition performance, the particle depositions and distributions of the two formulations on the rice leaf surface were observed using a scanning electron microscope. As shown in Figure 4C–E, there were abundant papillae and micro/nanostructures on the rice leaf surface, with widely distributed grooves. This microscopic morphology is more favorable for nanoparticle deposition. Figure 4F–K further confirm the above speculation. Chlorantraniliprole nanoparticles were easily distributed between the grooves, which could effectively inhibit the loss of active ingredients. In contrast, the large particles in the SC could not match the micro/nanostructures of the leaf surface, thus covering the top surface in a blocky form which can easily roll off. Therefore, the higher retention of the chlorantraniliprole nanosuspension can be attributed to its small size and enhanced foliar wettability, facilitating an improvement in the pesticide utilization rate and efficacy.

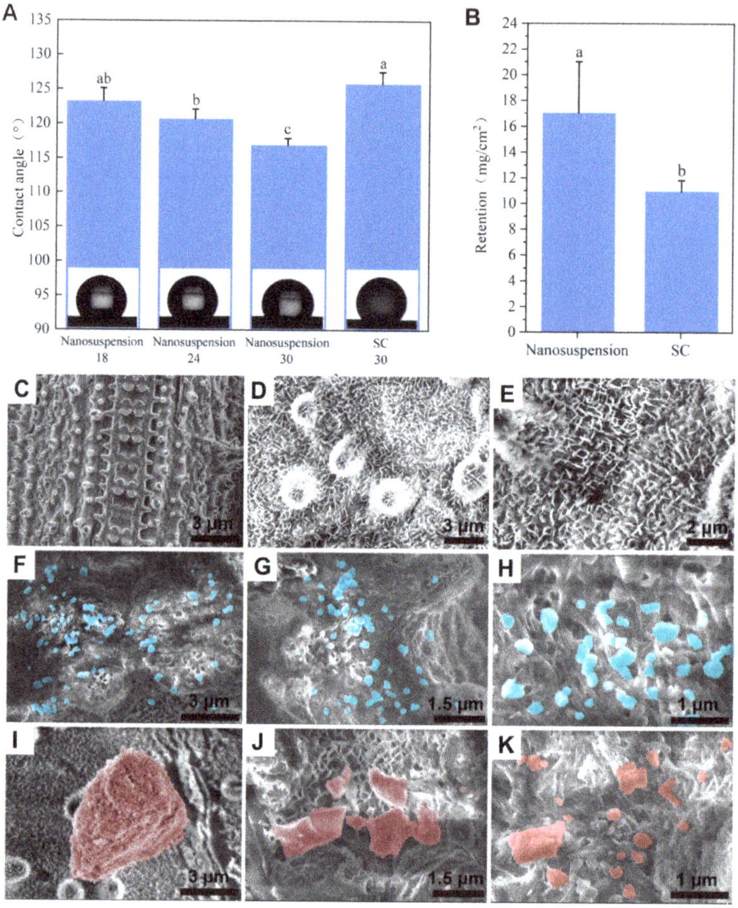

Figure 4. (**A**) Contact angles and (**B**) retention of the chlorantraniliprole nanosuspension and SC on rice leaves. (**C–E**) Morphology of blank rice leaves. Depositions and distributions of the chlorantraniliprole particles in (**F–H**) the nanosuspension and (**I–K**) the SC on rice leaves. Different letters (a,b,c) in the figure indicate significant differences at $p < 0.05$.

2.6. Field Efficacy on Cnaphalocrocis Medinalis

At present, knapsack sprayers still are the most common agricultural machinery used for pest control in China. However, in recent years, with the reduction in the rural workforce and a sharp increase in labor costs, coupled with the increase in agricultural intensification, plant protection drones have been widely used. Limited by low loading capacity, highly concentrated and dispersed pesticide formulations are more suitable for drone spraying. Hence, in this study, field efficacy evaluations of the chlorantraniliprole nanosuspension against *Cnaphalocrocis medinalis* were performed using two different spray modes: a knapsack sprayer and a drone. Under manual spraying, both the nanosuspension and the conventional SC achieved around 85% control efficacies against *Cnaphalocrocis medinalis* (Figure 5A). There was no significant difference in the control effect of the nanosuspension at 18 g a.i./hm^2 and SC at 30 g a.i./hm^2 dosages. A lower spraying height in manual mode reduced the influence of the external environment on pesticide droplet adhesion, thereby reducing the differences in efficacy between the various formulations. As shown in Figure 5B, when a drone was used for spraying, the nanosuspension efficacies were greater than 90% at application doses of 18–30 g a.i./hm^2, but the control effect of the SC was below 80%, which meant that the nanoformulation could reduce pesticide usage by about 40%. In high-concentration spraying, the better dispersibility and suspensibility of the nanosuspension were more conducive to improving the coverage, adhesion, and deposition of active ingredients on crop leaves, thus increasing the contact probability between pesticides and pests and further improving the biological control effect. Conversely, the large particles in the SC were more likely to agglomerate and settle, resulting in concentration discrepancies during the application process and inconsistent control against *Cnaphalocrocis medinalis*. It is evident that chlorantraniliprole nanoformulation use led to a substantial reduction in pesticide usage and increased the biocontrol efficiency, thereby increasing the formulation's ecological safety. This study provides a scientific basis and data-based support for the development and application of nanopesticides.

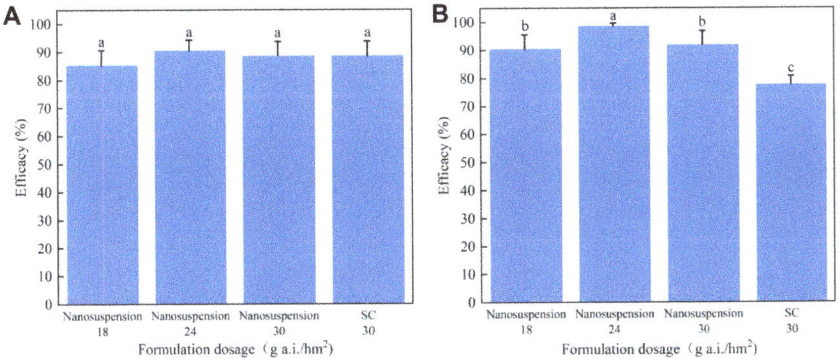

Figure 5. Field efficacies of the chlorantraniliprole nanosuspension and the SC against *Cnaphalocrocis medinalis* spraying using (**A**) a knapsack sprayer and (**B**) a drone. Different letters (a,b,c) in the figure indicate significant differences at $p < 0.05$.

2.7. Pesticide Residue

Pesticide residues in crop tissues and harvested fruits have a significant effect on the ecological environment and food safety. Table 4 compares the chlorantraniliprole residues on rice leaves and grains after applying the nanosuspension and the SC with the same active ingredient dosage. The maximum residue limits (MRLs) for chlorantraniliprole in rice vary according to countries and regions, with MRLs of 0.5 mg/kg, 0.4 mg/kg, 0.4 mg/kg, 0.02 mg/kg in China, Codex Alimentarius Commission (CAC), Australia, and Europe, respectively. The application of the chlorantraniliprole nanosuspension had low residue risks, thus ensuring food and ecosystem safety. Moreover, considering the field control efficacy, the chlorantraniliprole nanosuspension facilitated a reduction in pesticide usage, which further enhances the formulation's environmental friendliness, delays pest resistance, and improves the economic benefits.

Table 4. Residues of chlorantraniliprole in rice treated with the proposed nanosuspension and SC.

Sample	Treatment	Residue Amount (mg/kg)
Leaves	SC 30 g a.i./hm^2	0.074 ± 0.0021
	nanosuspension 30 g a.i./hm^2	0.063 ± 0.0006
Grains	SC 30 g a.i./hm^2	0.042 ± 0.0006
	nanosuspension 30 g a.i./hm^2	Not detected

2.8. Impact on Rice Quality

The content and distribution of major nutritional components in rice are pivotal indicators of rice quality. Table 5 shows the effects of the different chlorantraniliprole formulations on the nutrient content in rice grains. Compared to the control group, chlorantraniliprole application had no effect on protein content. Khan et al.'s study also showed that five insecticides—Lorsban (40% EC), Decis (25% EC), Pyrifos (40% EC), Karate (25% EC), and Ripcord (10% EC)—did not have any significant effect on the protein content of chickpea (*Cicer arietinum* L.) crops [54]. However, pesticide application reduced the contents of energy, fat, and carbohydrates, as well as calcium, magnesium, iron, and potassium elements in rice grains. Reddy et al. investigated the influence of insecticide spraying on mineral element contents in cabbage. They found that pesticides significantly reduced the zinc content but increased the calcium and potassium contents. Additionally, endosulfan had a more pronounced effect on altering mineral element contents compared to malathion [55]. The probable reason for this is that pesticides, as exotic stressors, cause a reduction in rice stomatal opening, which in turn results in a decrease in photosynthetic rate and organic matter production, thereby reducing the fat and carbohydrates stored in rice gains [56–60]. Khidir clarified that treflan could significantly reduce the amount of total nitrogen, phosphorus, potassium, sodium, iron, zinc, and manganese in wheat [61]. Moreover, the calcium, magnesium, iron, zinc, and potassium contents in gains treated with nanosuspension were all higher than those in the group sprayed with SC. The nanosuspension has a smaller and more uniform particle size, and may have a lower impact on stomatal opening and photosynthesis. Rice transports mineral elements to plant tissues via the chelation of deoxymalturonic acid (DMA) and mineral elements [62]. Pesticides lost to the soil may also affect the dissolution and chelation of mineral elements, thereby affecting their delivery [63]. In conclusion, the nanosuspension has a smaller impact on the quality of rice.

Table 5. Effects of applying the chlorantraniliprole nanosuspension and the SC on rice quality.

Nutrient	Control Group	Nanosuspension	SC
Energy	1550 kJ/100 g	1511 kJ/100 g	1498 kJ/100 g
Protein	7.2 g/100 g	7.2 g/100 g	7.2 g/100 g
Fat	1.4 g/100 g	0.9 g/100 g	0.9 g/100 g
Carbohydrate	80.0 g/100 g	78.6 g/100 g	77.8 g/100 g
Dietary fiber	2.0 g/100 g	2.4 g/100 g	2.5 g/100 g
Calcium	89.4 mg/kg	88.2 mg/kg	81.7 mg/kg
Magnesium	587 mg/kg	556 mg/kg	413 mg/kg
Iron	7.82 mg/kg	6.21 mg/kg	5.41 mg/kg
Zinc	21.4 mg/kg	23.1 mg/kg	19.6 mg/kg
Potassium	1630 mg/kg	1480 mg/kg	1170 mg/kg

2.9. Zebrafish Safety

Pesticides can enter the aquatic environment of rice paddies and cause serious ecological and environmental problems. In this study, the safety of chlorantraniliprole regarding aquatic organisms was evaluated using zebrafish as a model species. As shown in Figure 6, the mortality of zebrafish increased with increasing chlorantraniliprole concentrations and treatment durations. The LC_{50} value of the chlorantraniliprole SC was 69.13 mg/L, consistent with what is described in [64]. In contrast, all zebrafish survived at each corresponding concentration in the chlorantraniliprole nanosuspension group, even at 500 mg/L. Considering that it would be difficult to reach such high chlorantraniliprole concentrations in the natural environment, it can be inferred that the nanosuspension is extremely safe for zebrafish. An analogous phenomenon has also been noted by other researchers. Huang et al. compared the acute toxicities of three chlorantraniliprole formulations to zebrafish embryos. The LC_{50} of the aqueous solution was greater than 80 mg/L, while the LC_{50} values of the SC and granules were 32.34 mg/L and 25.96 mg/L, respectively [65]. The significant difference in toxicity to zebrafish between the different formulations is mainly linked to the type and content of organic solvents and surfactants in the compositions. In this study, the low toxicity of the chlorantraniliprole nanosuspension was due to chlorantraniliprole's action mechanism, which interferes with the nervous system of pests to cause paralysis or death [66]. Essentially, chlorantraniliprole is an insensitive agent to zebrafish. In particular, the processing stages and the nanosuspension composition are free of organic solvents and harmful adjuvants, thus rendering it environmentally friendly and safe for non-target organisms.

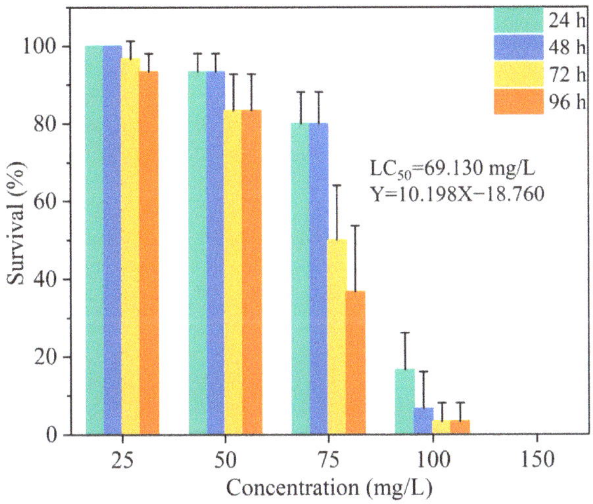

Figure 6. Toxicity of the chlorantraniliprole SC to zebrafish.

3. Materials and Methods

3.1. Materials

The chlorantraniliprole technical material (TC, 95%, w/w) and suspension concentrate (SC, 200 g/L) were acquired from FMC Corporation (Shanghai, China). Maleic rosinpolyoxypropylene-polyoxyethylene ether sulfonate (MRES) and polycarboxylate were provided by Sinvochem S&D Co., Ltd. (Yangzhou, Jiangsu, China). Chromatographic-grade acetonitrile was bought from Thermo Fisher Scientific Co., Ltd. (Shanghai, China). Milli-Q water (18 MΩ cm, TOC $\leq$ 4 ppb) was used in all analytical experiments.

3.2. Preparation of Chlorantraniliprole Nanosuspension

In our previous work, it was proven that a composite surfactant consisting of MRES and polycarboxylate (1:1, w/w) is a suitable surfactant system for the prevention of chlorantraniliprole nanoparticle aggregation [67]. Based on this, the same surfactant combination was applied to stabilize a chlorantraniliprole nanosuspension produced via the wet media milling method. The specific preparation procedure was as follows. Firstly, 5.62 g of MRES and the same amount of polycarboxylate were dissolved in 154 mL of water to obtain a colorless and transparent solution. Subsequently, 22.5 g of chlorantraniliprole was added into the above solution, and the mixture was stirred and emulsified at rcf = 1453 g for 10 min in a shearing machine (25BC, Shanghai HENC Mechanical Equipment Co., Ltd., Shanghai, China) to uniformly suspend the pesticide particles in the suspension. Then, the dispersion was transferred to laboratory milling equipment (WG-0.3, Suzhou Vgreen Nano-Chem Technology Co., Ltd., Suzhou, Jiangsu, China) to produce chlorantraniliprole nanosuspensions. The milling media were 0.3 mm zirconium dioxide beads. The surfactant content and milling time changed with the experimental design.

3.3. Particle Size and Zeta Potential Measurements

The chlorantraniliprole nanosuspension and SC were diluted with water and then placed into the sample cells of a Zetasizer Nano ZS 90 (Malvern, UK). The particle size, polydispersity index (PDI), and zeta potential were measured at room temperature. Each sample was measured in triplicate, and the data were recorded as means ± standard deviation (S.D.).

3.4. Morphological Characterization

The morphological characterization of the chlorantraniliprole particles was performed using a scanning electron microscope (JSM-7401F, JEOL, Tokyo, Japan). Aqueous dispersions of the chlorantraniliprole nanosuspension and SC were dropped onto freshly cleaned silicon slices. The samples were air-dried and coated with platinum using a sputter coater (ETD-800, Beijing Elaborate Technology Development Ltd., Beijing, China). The images were recorded in low electron image (LEI) mode, and the statistical particle sizes were determined based on SEM images using Nano Measurer software (1.2.5).

3.5. Stability Test

The physical stability of the chlorantraniliprole nanosuspension was assessed according to CIPAC MT 39, CIPAC MT 46, and GB/T 19136–2003. The samples were stored at 0 ± 2 °C, 25 ± 2 °C, and 54 ± 2 °C for 14 days. The particle size and PDI were measured every two days.

3.6. Contact Angle Measurement

The contact angles of the chlorantraniliprole nanosuspension and SC on rice leaves were measured using a contact angle apparatus (JC2000D, Zhongchen Digital Technic Apparatus Co., Ltd., Shanghai, China). The samples were diluted with water to different concentrations corresponding to those applied in the field. Fresh rice leaves were smoothly adhered to glass slides. Then, a drop of the sample aqueous dispersion was placed onto the surface of the rice leaf. The droplet was photographed after standing for 15 s, and the five-point fitting method was used to calculate the contact angle. The average value of five replicates was calculated.

3.7. Retention Test

The retention was measured (R_m, mg/cm) according to the method described in [68] with slight modifications. First, the chlorantraniliprole nanosuspension and SC were diluted into aqueous dispersions containing 0.1% (*w*/*w*) of active ingredient. Secondly, each leaf was weighed using an electronic balance (ME204E, Mettler Toledo, Zurich, Switzerland), and its surface area was measured using a leaf area meter (Yaxin-1241, Beijing Yaxin Science Instrument Technology Co., Ltd., Beijing, China). Leaves were then completely immersed in the above dispersions. After 15 s, each leaf was removed and weighed again. The retention (R_m) was calculated based on the following equation. The particle adsorption and distribution of the chlorantraniliprole nanosuspension and SC on rice leaves were also observed using a scanning electron microscope.

$$R = \frac{M_1 - M_0}{S} \tag{1}$$

where M_0 (mg) and M_1 (mg) are the leaf weights before and after immersion in dispersions, and S (cm^2) is the leaf area. The average of five tests was calculated.

3.8. Field Efficacy on Cnaphalocrocis Medinalis

The field efficacies of the chlorantraniliprole nanosuspension and SC with different size characteristics were compared. The field trials for the control effect on *Cnaphalocrocis medinalis* were performed at two sites. A knapsack sprayer was used to spray pesticide in Baiertu Village, Lubu Town, Zhaoqing City, Guangdong Province, China, from 10 to 25 May 2019, while a drone was applied for spraying at Mawang Village, Wuling Town, Bingyang County, Nanning City, Guangxi Province, China, from 2 to 17 June 2019. Each experimental area was 667 m^2 and arranged in a randomized block design. The concentration of the chlorantraniliprole SC was set as 30 g a.i./hm^2, considering the recommended dosage. The nanosuspension was diluted into three concentration gradients of 18, 24, and 30 g a.i./hm^2. The parallel skip method was used for sampling, and 50 samples were taken from each treatment area. Three replicates were conducted for each treatment.

3.9. *Residue Test*

3.9.1. Sample Collection and Processing

Rice leaf samples were collected 21 days after pesticide application, and grain samples were collected after the rice was mature. Samples were gathered independently from at least 12 random sites in each plot. After collection, 2.000 g of accurately weighed rice leaves (cutting into pieces less than 1 cm) or crushed grains was placed into 50 mL centrifuge tubes. Subsequently, 19.8 mL of acetonitrile, 0.2 mL of acetic acid, 1.5 g of anhydrous sodium acetate, and 6 g of anhydrous magnesium sulfate were added and homogenized at RCF = 503 g for 1 min, followed by centrifugation at RCF = 905 g for 5 min. Then, 6 mL of the upper acetonitrile extract was added into a centrifuge tube filled with 900 mg of anhydrous magnesium sulfate and 150 mg of PSA (primary secondary amine) sorbent. After vortex mixing for 1 min and centrifugation at RCF = 905 g for 5 min, 0.75 mL of the supernatant was mixed with water in a 1:1 ratio (v/v) and passed through a 0.22 μm filter membrane for testing. The same method was used to determine the chlorantraniliprole recovery rate. Three concentration levels (10 μg/L, 20 μg/L, and 30 μg/L) were used for the recovery test, and three parallel tests were conducted for each sample.

3.9.2. Chromatographic Analysis Conditions

Liquid phase separation was performed by gradient elution using a Kinetex C18 column (4.6 mm × 100 mm, 2.6 μm) under the conditions listed in Table 6. The analysis time for each sample was 7 min. The mobile phase consisted of water (with 1‰ formic acid) and acetonitrile. The column temperature was maintained at 25 °C.

Table 6. Mobile phase parameters for gradient elution in liquid chromatography.

Time (min)	Flow Rate (mL/min)	Water Phase	Acetonitrile
0	0.6	60	40
0.5	0.6	60	40
5	0.6	5	95
5.5	0.6	5	95
6	0.6	60	40
7	0.6	60	40

3.9.3. Mass Spectrometric Analysis Conditions

Chlorantraniliprole was detected in positive-ion mode because of the minimal background interferences and high response. The protonated molecular ion peak $[M + H]^+$ of chlorantraniliprole was observed at m/z 484. The $[M + H]^+$ ion was selected as the precursor ion for secondary mass spectrometry analysis, and the major fragmentation ions of chlorantraniliprole were located at m/z 453 and 286. These fragmentation ions exhibited strong and stable responses, thus acting as product ions. An automated optimization approach was employed to maximize the response of the target compound under multiple reaction monitoring (MRM) mode, ensuring the highest sensitivity to analytes.

3.9.4. Measurement of Additive Recovery

The standard curve equation of chlorantraniliprole, measured by liquid chromatography–mass spectrometry, was Y = 8020.2X + 835.67, with a correlation coefficient R2 of 0.9999. This indicates that there was a good linear relationship between the concentration of chlorantraniliprole and the peak area in the range of 0.5–50 μg/L. A recovery test at three chlorantraniliprole concentrations was carried out, and the average recovery rate of the added standard concentrations of 10 μg/L, 20 μg/L, and 30 μg/L ranged from 80.1 to 109.63%. The results in

Table 7 show that the RSD of the three measurements is 0.9–4.2%, indicating good recovery and reproducibility.

Table 7. Recovery rate of chlorantraniliprole addition in rice.

Treatment	Number	10 μg/L	20 μg/L	30 μg/L	RSD (%)
Chlorantraniliprole (leaf)	1	79.6%	89.7%	109.2%	0.9
	2	78.4%	87.6%	107.4%	
	3	82.3%	85.4%	112.3%	
	average	80.10%	87.57%	109.63%	
Chlorantraniliprole (grain)	1	80.2%	85.6%	105.9%	4.2
	2	78.9%	84.3%	107.3%	
	3	84.1%	80.1%	99.6%	
	average	81.07%	83.33%	104.20%	

3.10. Rice Quality Determination

The protein content in rice was determined using the Kjeldahl nitrogen method according to GB 5009.5-2016. In detail, 3 g of rice was taken, and under catalytic heating conditions, proteins were decomposed, and the produced ammonia was combined with sulfuric acid to form ammonium sulfate. Alkaline distillation was performed to release ammonia, which was then absorbed by boric acid and titrated using a standard sulfuric acid solution. The nitrogen content was calculated based on the consumption of acid and multiplied by the conversion factor to obtain the protein content. The determination of fat in rice was carried out using the Soxhlet extraction method following GB 5009.6-2016. For this, 3 g of rice was accurately weighed and placed in a filter paper tube. The tube was then put into a Soxhlet extractor with anhydrous ether, and the mixture was heated under reflux. After drying, the fat mass was measured. The calculation of the energy and total carbohydrates in rice was based on GB 28050-2011. The total carbohydrates were calculated as the total of food mass minus the mass of protein, fat, water, and ash. The energy content was calculated by multiplying the contents of protein, fat, and total carbohydrates by energy factors of 17, 37, and 17 kJ/g, respectively, and then summing them up per 100 g of the product. The determination of mineral trace elements in rice was carried out using the Inductively Coupled Plasma Atomic Emission Spectrometry (ICP-AES) method following GB 5009.268-2016. After the 3 g rice sample was digested, it was measured using an inductively coupled plasma atomic emission spectrometer. The qualitative analysis was based on the characteristic spectral line wavelength of the element, and the quantitative analysis was performed using the direct proportionality between the signal intensity of the element spectral line and its concentration.

3.11. Toxicity against Zebrafish

The chlorantraniliprole nanosuspension and SC were diluted with water to obtain sample suspensions at several different concentrations. Ten adult zebrafish were randomly selected and transferred to the sample suspensions. During the period of the experiment, the suspensions were replaced every 24 h to maintain the pesticide concentration and water quality. All zebrafish were reared at a temperature of 25 ± 1 °C. The poisoning symptoms and mortality of the tested zebrafish within 24, 48, 72, and 96 h were observed. The judgment standard for death status was no breathing or no movement upon touching the tail.

3.12. Statistical Analysis

Data were analyzed using a one-way analysis of variance (ANOVA) and Duncan's multiple range tests, and a significance level of less than 0.05 was considered statistically significant.

4. Conclusions

In this study, a chlorantraniliprole nanosuspension with an average particle size of 56 nm and excellent stability was produced by using wet media milling technology, providing a green and scalable strategy to construct nano-delivery systems for pesticides that are poorly soluble in both water and organic solvents. Compared to the SC formulation, the nanosuspension exhibited better dispersibility, foliar wetting, and retention. The field trials confirmed that pesticide usage could be reduced by about 40% through using the nanoformulation while maintaining the same level of efficacy as the SC. It is worth noting that the nanosuspension exhibited lower residual properties and higher non-target biosafety while exerting a high insecticidal activity, reducing the negative impact of pesticides on rice and ensuring rice quality and food safety. These advantages are attributed to the nanosuspension's environmentally friendly composition and processing procedure. Comprehensive efficacy and safety evaluations of chlorantraniliprole nanosuspensions are of paramount importance for guiding the design, construction, and application of nanopesticides. This highly effective and eco-friendly nanosuspension has broad application prospects in crop protection for improving pesticide efficacy and reducing residual pollution in agricultural products and the environment.

Author Contributions: Conceptualization and methodology, B.C., F.G. and H.C.; data analysis, X.D., L.G. and Q.D.; writing—original draft preparation, X.D. and T.W.; writing—review and editing, B.C., F.G., Y.W. and Z.Z.; funding acquisition, B.C. All authors have read and agreed to the published version of the manuscript.

Funding: This work was financially supported by the National Natural Science Foundation of China (22278423) and the Key Technologies R&D Program of Guangdong Province (2023B0202080002).

Institutional Review Board Statement: Not applicable.

Informed Consent Statement: Not applicable.

Data Availability Statement: Research data are available upon request to the authors.

Conflicts of Interest: The authors declare no conflict of interest.

References

1. Cordova, D.; Benner, E.A.; Sacher, M.D.; Rauh, J.J.; Sopa, J.S.; Lahm, G.P.; Selby, T.P.; Stevenson, T.M.; Flexner, L.; Gutteridge, S. Anthranilic diamides: A new class of insecticides with a novel mode of action, ryanodine receptor activation. *Pestic. Biochem. Physiol.* **2006**, *84*, 196–214. [CrossRef]
2. Cordova, D.; Benner, E.A.; Sacher, M.D.; Rauh, J.J.; Sopa, J.S.; Lahm, G.P.; Selby, T.P.; Stevenson, T.M.; Flexner, L.; Caspar, T. Elucidation of the Mode of Action of Rynaxypyr^{TM}, a Selective Ryanodine Receptor Activator. In *Pesticide Chemistry: Crop Protection, Public Health, Environmental Safety*; Wiley: Hoboken, NJ, USA, 2007; pp. 121–126.
3. Sial, A.A.; Brunner, J.F.; Garczynski, S.F. Biochemical characterization of chlorantraniliprole and spinetoram resistance in laboratory-selected obliquebanded leafroller, Choristoneura rosaceana (Harris) (Lepidoptera: Tortricidae). *Pestic. Biochem. Physiol.* **2011**, *99*, 274–279. [CrossRef]
4. Lu, Y.; Wang, G.; Zhong, L.; Zhang, F.; Bai, Q.; Zheng, X.; Lu, Z. Resistance monitoring of Chilo suppressalis (Walker) (Lepidoptera: Crambidae) to chlorantraniliprole in eight field populations from east and central China. *Crop. Protect.* **2017**, *100*, 196–202. [CrossRef]
5. Son, M.; Ju, J.; Lu, S.; Han, Y.; Don, Z.; Wan, Y.; Zhen, G.; Zhan, L.; Ha, R.; Jiang, L. Controlling liquid splash on superhydrophobic surfaces by a vesicle surfactant. *Sci. Adv.* **2017**, *3*, e1602188.
6. Nuruzzaman, M.; Rahman, M.M.; Liu, Y.; Naidu, R. Nanoencapsulation, nano-guard for pesticides: A new window for safe application. *J. Agric. Food. Chem.* **2016**, *64*, 1447–1483. [CrossRef] [PubMed]
7. Gan, J.; Hussain, M.; Rathor, N.M. Behaviour of an alginate-kaolin based controlled-release formulation of the herbicide thiobencarb in simulated ecosystems. *Pestic. Sci.* **1994**, *42*, 265–272. [CrossRef]
8. Mogul, M.G.; Akin, H.; Hasirci, N.; Trantolo, D.J.; Gresser, J.D.; Wise, D.L. Controlled release of biologically active agents for purposes of agricultural crop management. *Resour. Conserv. Recycl.* **1996**, *16*, 289–320. [CrossRef]
9. Margni, M.D.P.O. Life cycle impact assessment of pesticides on human health and ecosystems. *Agric. Ecosyst. Environ.* **2002**, *93*, 379–392. [CrossRef]
10. Gong, C.W.; Hasnain, A.; Wang, Q.L.; Liu, D.; Xu, Z.Z.; Zhan, X.X.; Liu, X.M.; Pu, J.; Sun, M.M.; Wang, X.G. Eco-friendly deacetylated chitosan base siRNA biological-nanopesticide loading cyromazine for efficiently controlling Spodoptera frugiperda. *Int. J. Biol. Macromol.* **2023**, *241*, 124575. [CrossRef]

11. Song, S.J.; Wan, M.H.; Feng, W.L.; Tian, Y.; Jiang, X.F.; Luo, Y.; Shen, J. Environmentally friendly Zr-based MOF for pesticide delivery: Ultrahigh loading capacity, pH-responsive release, improved leaf affinity, and enhanced antipest activity. *Langmuir* **2022**, *38*, 10867–10874. [CrossRef]
12. Zhou, Y.; Wu, J.Y.Z.; Zhou, J.; Lin, S.K.; Cheng, D.M. pH-responsive release and washout resistance of chitosan-based nano-pesticides for sustainable control of plumeria rust. *Int. J. Biol. Macromol.* **2022**, *222*, 188–197. [CrossRef]
13. Rai, M.; Ingle, A. Role of nanotechnology in agriculture with special reference to management of insect pests. *Appl. Microbiol. Biotechnol.* **2012**, *94*, 287–293. [CrossRef] [PubMed]
14. Khan, S.; Zahoor, M.; Khan, R.S.; Ikram, M.; Ul Islam, N. The impact of silver nanoparticles on the growth of plants: The agriculture applications. *Heliyon* **2023**, *9*, e16928. [CrossRef] [PubMed]
15. Zhang, X.; Xia, Q.; Gu, N. Preparation of all-trans retinoic acid nanosuspensions using a modified precipitation method. *Drug Dev. Ind. Pharm.* **2006**, *32*, 857. [CrossRef] [PubMed]
16. Pandey, N.K.; Garg, V.; Bhattacharya, S.; Gulati, M.; Vaidya, Y.; Singh, S.K. Nanosuspension: Principles, perspectives and practices. *Curr. Drug Del.* **2016**, *13*, 1222–1246.
17. Zhang, D.; Tan, T.; Gao, L.; Zhao, W.; Wang, P. Preparation of azithromycin nanosuspensions by high pressure homogenization and its physicochemical characteristics studies. *Drug Dev. Ind. Pharm.* **2007**, *33*, 569–575. [CrossRef] [PubMed]
18. Xiong, R.; Lu, W.; Li, J.; Wang, P.; Xu, R.; Chen, T. Preparation and characterization of intravenously injectable nimodipine nanosuspension. *Int. J. Pharm.* **2008**, *350*, 338–343. [CrossRef]
19. MoSchwitzer, J.; Achleitner, G.; Pomper, H.; Müller, R.H. Development of an intravenously injectable chemically stable aqueous omeprazole formulation using nanosuspension technology. *Eur. J. Pharm. Biopharm.* **2004**, *58*, 615–619. [CrossRef]
20. Muller, R.H.M.; Peters, K. Nanosuspensions for the formulation of poorly soluble drugs. *Int. J. Pharm.* **1998**, *160*, 229–237. [CrossRef]
21. Eerdenbrugh, B.V.; Mooter, G.V.D.; Augustijns, P. Top-down production of drug nanocrystals: Nanosuspension stabilization, miniaturization and transformation into solid products. *Int. J. Pharm.* **2008**, *364*, 64–75. [CrossRef]
22. Chingunpitak, J.; Puttipipatkhachorn, S.; Chavalitshewinkoon-Petmitr, P.; Tozuka, Y.; Moribe, K.; Yamamoto, K. Formation, physical stability and in vitro antimalarial activity of dihydroartemisinin nanosuspensions obtained by co-grinding method. *Drug Dev. Ind. Pharm.* **2008**, *34*, 314–322. [CrossRef] [PubMed]
23. Agrawal, Y.; Patel, V. Nanosuspension: An approach to enhance solubility of drugs. *J. Adv. Pharm. Technol. Res.* **2011**, *2*, 81–87. [CrossRef]
24. Chin, C.-P.; Wu, H.-S.; Wang, S.S. New approach to pesticide delivery using nanosuspensions: Research and applications. *Ind. Eng. Chem. Res.* **2011**, *50*, 7637–7643. [CrossRef]
25. Corrias, F.; Melis, A.; Atzei, A.; Marceddu, S.; Angioni, A. Zoxamide accumulation and retention evaluation after nanosuspension technology application in tomato plant. *Pest Manag. Sci.* **2021**, *77*, 3508–3518. [CrossRef]
26. Cui, B.; Lv, Y.; Gao, F.; Wang, C.; Zeng, Z.; Wang, Y.; Sun, C.; Zhao, X.; Shen, Y.; Liu, G. Improving abamectin bioavailability via nanosuspension constructed by wet milling technique. *Pest Manag. Sci.* **2019**, *75*, 2756–2764. [CrossRef] [PubMed]
27. Zhenzhong, P.; Bo, C.; Zhanghua, Z.; Lei, F.; Guoqiang, L.; Haixin, C.; Hongyu, P. Lambda-Cyhalothrin nanosuspension prepared by the melt emulsification-high pressure homogenization method. *J. Nanomater.* **2015**, *16*, 263.
28. Wang, C.; Cui, B.; Guo, L.; Wang, A.; Zhao, X.; Wang, Y.; Sun, C.; Zeng, Z.; Zhi, H.; Chen, H. Fabrication and evaluation of Lambda-Cyhalothrin nanosuspension by one-Step melt emulsification technique. *Nanomaterials* **2019**, *9*, 145. [CrossRef]
29. Rani, L.; Thapa, K.; Kanojia, N.; Sharma, N.; Singh, S.; Grewal, A.S.; Srivastav, A.L.; Kaushal, J. An extensive review on the consequences of chemical pesticides on human health and environment. *J. Clean. Prod.* **2021**, *283*, 124657. [CrossRef]
30. Yang, B. The impact of pesticide residues on food safety. *China Food Saf. Mag.* **2021**, *22*, 152–153.
31. Sumira, M.; Shilpa, P.; Shristi, K.; Abhishek, K.; Vineet, U. A perspective review on impact and molecular mechanism of environmental carcinogens on human health. *Biotechnol. Genet. Eng. Rev.* **2021**, *37*, 178–207.
32. Chen, L.; Jiang, H.; Zhou, Y.; Zhou, X.; Huang, J. A Brief Analysis on the Environmental Safety of Nano-enabled Pesticides. *Pestic. Sci. Adm.* **2018**, *39*, 9.
33. Sun, Y.; Liang, J.; Tank, L.; Li, H.; Zhu, Y.; Jiang, D.; Song, B.; Chen, M.; Zeng, G. Nano-pesticides: A great challenge for biodiversity? *Nano Today* **2019**, *28*, 100757. [CrossRef]
34. Berger, S.; Cwiek, K. Selected aspects of adverse nutritional effects of pesticides. *Ernhrung* **1990**, *14*, 411–415.
35. Rico, C.M.; Morales, M.I.; Barrios, A.C.; McCreary, R.; Hong, J.; Lee, W.-Y.; Nunez, J. Effect of cerium oxide nanoparticles on the quality of rice (*Oryza sativa* L.) grains. *J. Agric. Food. Chem.* **2013**, *61*, 11278–11285. [CrossRef] [PubMed]
36. Zhao, P.; Wang, C.; Zhang, S.; Zheng, L.; Li, F.; Cao, C.; Cao, L.; Huang, Q. Fungicide-loaded mesoporous silica nanoparticles promote rice seedling growth by regulating amino acid metabolic pathways. *J. Hazard. Mater.* **2022**, *425*, 127892. [CrossRef] [PubMed]
37. Cui, B.; Gao, F.; Zeng, Z.H.; Wang, C.X.; Wang, Y.; Sun, C.J.; Zhao, X.; Guo, L.; Shen, Y.; Liu, G.Q.; et al. Construction and characterization of avermectin B-2 solid nanodispersion. *Sci. Rep.* **2020**, *10*, 9096. [CrossRef]
38. Danaei, M.; Dehghankhold, M.; Ataei, S.; Hasanzadeh, D.F.; Javanmard, R.; Dokhani, A.; Khorasani, S.; Mozafari, M. Impact of particle size and polydispersity index on the clinical applications of lipidic nanocarrier systems. *Pharmaceutics* **2018**, *10*, 57. [CrossRef]

39. Lipinski, C.A. Drug-like properties and the causes of poor solubility and poor permeability. *J. Pharmacol. Toxicol. Methods* **2000**, *44*, 235–249. [CrossRef]
40. Kawakami, K. Modification of physicochemical characteristics of active pharmaceutical ingredients and application of supersaturatable dosage forms for improving bioavailability of poorly absorbed drugs. *Adv. Drug Deliv. Rev.* **2012**, *64*, 480–495. [CrossRef]
41. Kipp, J.E. The role of solid nanoparticle technology in the parenteral delivery of poorly water-soluble drugs. *Int. J. Pharm.* **2004**, *284*, 109–122. [CrossRef]
42. Gittings, M.R.; Saville, D.A. The determination of hydrodynamic size and zeta potential from electrophoretic mobility and light scattering measurements. *Colloids Surf. A Physicochem. Eng. Asp.* **1998**, *141*, 111–117. [CrossRef]
43. Tapak, N.S.; Nawawi, M.A.; Mohamed, A.H.; Tjih, E.T.T.; Mohd, Y.; Ab Rashid, A.H.B.; Abdullah, J.; Yusof, N.A.; Ahmad, N.M. Chemical synthesis of metal oxide nanoparticles via ionic liquid as capping agent: Principle, preparation and applications. *Malays. J. Anal. Sci.* **2022**, *26*, 1394–1420.
44. Feng, J.; Lu, F.; Li, M.; Li, W.; Wang, X. Stability of suspension solution and development of suspension concentrate products. *Agric. Res. Appl.* **2009**, *3*, 12–19.
45. Chunxin, W.; Bo, C.; Xiang, Z.; Yan, W.; Zhanghua, Z.; Changjiao, S.; Dongsheng, Y.; Guoqiang, L.; Haixin, C. Optimization and characterization of lambda-cyhalothrin solid nanodispersion by self-dispersing method. *Pest Manag. Sci.* **2018**, *75*, 380–389.
46. Liu, H.H.; Surawanvijit, S.; Rallo, R.; Orkoulas, G.; Cohen, Y. Analysis of nanoparticle agglomeration in aqueous suspensions via constant-number Monte Carlo simulation. *Environ. Sci. Technol.* **2011**, *45*, 9284–9292. [CrossRef]
47. Bhattacharjee, S. DLS and zeta potential–what they are and what they are not? *J. Control. Release* **2016**, *235*, 337–351. [CrossRef] [PubMed]
48. Müller, K.H.; Motskin, M.; Philpott, A.J.; Routh, A.F.; Shanahan, C.M.; Duer, M.J.; Skepper, J.N. The effect of particle agglomeration on the formation of a surface-connected compartment induced by hydroxyapatite nanoparticles in human monocyte-derived macrophages. *Biomaterials* **2014**, *35*, 1074–1088. [CrossRef] [PubMed]
49. Farrell, E.; Brousseau, J.-L. Guide for DLS sample preparation. *Brookhaven. Instrum.* **2014**, *1*, 1–3.
50. Ma, H.; Xia, S.; Li, N.; Wang, T.; Zheng, W.; Yu, T.; Shu, Q.; Han, Y. Emulsifying stability and viscosity reduction for heavy crude oil in surfactant-polymer composite system. *J. Mol. Liq.* **2022**, *362*, 119713. [CrossRef]
51. Zhang, W.; Wang, Y.; Wang, S.; Guo, Z.; Zhang, C.; Zhu, X.; Zhang, G. Hyperbranched ionic surfactants with polyether skeleton: Synthesis, properties and used as stabilizer for emulsion polymerization. *J. Mol. Liq.* **2022**, *355*, 118937. [CrossRef]
52. Bao, Z.; Wu, Y.; Liu, R.; Zhang, S.; Chen, Y.; Wu, T.; Gao, Y.; Zhang, C.; Du, F. Molecular selection and environmental evaluation of eco-friendly surfactants to efficiently reduce pesticide pollution. *J. Clean. Prod.* **2023**, *416*, 137954. [CrossRef]
53. He, J.; Li, J.; Gao, Y.; He, X.; Hao, G. Nano-based smart formulations: A potential solution to the hazardous effects of pesticide on the environment. *J. Hazard. Mater.* **2023**, *456*, 131599. [CrossRef] [PubMed]
54. Khan, H.; Zeb, A.; Ali, Z.; Shah, S.M. Impact of five insecticides on chickpea (*Cicer arietinum* L.) nodulation, yield and nitrogen fixing rhizospheric bacteria. *Soil Environ.* **2009**, *28*, 56–59.
55. Reddy, N.S.; Dashsontakke, S. Effect of spraying selected pesticides on the contents of specified minerals in cabbage. *Plant Foods Hum. Nutr. (Former. Qual. Plant.)* **1997**, *51*, 357–363. [CrossRef]
56. Ferree, D.C. Influence of Pesticides on Photosynthesis of Crop Plants. In Proceedings of the Photosynthesis and Plant Development, Diepenbeek, Belgium, 23–29 July 1978.
57. Hu, J. *Effects of Chemical Pesticide on Physiology and Biochemistry of Rice Plant and Rice Quality and Analysis of Residues*; Yangzhou University: Yangzhou, China, 2008.
58. Luo, S.; Wang, Z.; Feng, X.; Xu, J.; Ding, H.; Wu, J.; Ge, C.; Ma, F. Study on tracer dynamics of effects of pesticides on export rate of photosynthate of rice leaves. *Sci. Agric. Sin.* **2002**, *35*, 5.
59. Yuan, S.; Wu, J.; Xu, J.; Li, G. Influences of herbicides on physiology and biochemistry of rice. *Acta Phytophylacica Sin.* **2001**, *28*, 5.
60. Giménez-Moolhuyzen, M.; van der Blom, J.; Lorenzo-Mínguez, P.; Cabello, T.; Crisol-Martínez, E. Photosynthesis inhibiting effects of pesticides on sweet pepper leaves. *Insects* **2020**, *11*, 69. [CrossRef]
61. Khidir, S.M. Effect of some soil treated pesticides on growth characteristics of faba bean and wheat plants. *Int. Journ. Emerg. Techn. Comput. Appl. Sci.* **2013**, *5*, 7–20.
62. Bashir, K.; Seki, M.; Nishizawa, N.K. The transport of essential micronutrients in rice. *Mol. Breed.* **2019**, *39*, 168. [CrossRef]
63. Scheepmaker, J.W.A.; Kassteele, J.V.D. Effects of chemical control agents and microbial biocontrol agents on numbers of non-target microbial soil organisms: A meta-analysis. *Biocontrol. Sci. Technol.* **2011**, *21*, 1225–1242. [CrossRef]
64. Ding, J.; Liu, J. Acute toxicity test of 200 g/L chloride worm benzamide suspension agent against zebrafish. *J. Agric. Catastrophology* **2022**, *12*, 43–45.
65. Huang, W. *Toxcity Study of Two Kinds of Diamide Insecticides to Zebrafish Embryo*; Hainan University: Haikou, China, 2017.
66. Du, J.; Fu, Y. Diamide insecticides targeting insect ryanodine receptors: Mechanism and application prospect. *Biochem. Biophys. Res. Commun.* **2023**, *670*, 19–26. [CrossRef] [PubMed]

67. Cui, B.; Feng, L.; Wang, C.; Yang, D.; Cui, H. Stability and biological activity evaluation of chlorantraniliprole solid nanodispersions prepared by high pressure homogenization. *PLoS ONE* **2016**, *11*, e0160877. [CrossRef] [PubMed]
68. Yuan, H.; Qi, S.; Yang, D. Study on the point of run-off and the maximum retention of spray liquid on crop leaves. *Chin. J. Pestic. Sci.* **2000**, *2*, 66–71.

toxics

MDPI

Article

Ultrastructural Changes in the Midgut of Brazilian Native Stingless Bee *Melipona scutellaris* Exposed to Fungicide Pyraclostrobin

Caio E. C. Domingues [1,2,*], Lais V. B. Inoue [2], Aleš Gregorc [1], Leticia S. Ansaloni [1], Osmar Malaspina [2] and Elaine C. Mathias da Silva [3]

1 Faculty of Agriculture and Life Sciences, University of Maribor, Pivola 10, 2311 Hoče, Slovenia; leticia.salvioni1@um.si (L.S.A.)

2 Centro de Estudos de Insetos Sociais (CEIS), Departamento de Biologia Geral e Aplicada, Instituto de Biociências (IB), Universidade Estadual Paulista (UNESP)—"Júlio de Mesquita Filho", Rio Claro 13506-900, SP, Brazil; lais.inoue@unesp.br (L.V.B.I.); osmar.malaspina@unesp.br (O.M.)

3 Laboratório de Ecotoxicologia e Análise de Integridade Ambiental (LEIA), Departamento de Biologia (DBio), Universidade Federal de São Carlos (UFSCar), Sorocaba 18052-780, SP, Brazil; elaine@ufscar.br

* Correspondence: cecdomingues@gmail.com

Abstract: *Melipona scutellaris* is a Brazilian stingless bee that is important for pollinating wild flora and agriculture crops. Fungicides have been widely used in agriculture, and floral residues can affect forager bees. The goal of our study was to evaluate the effects of sublethal concentrations of pyraclostrobin on the midgut ultrastructure of *M. scutellaris* forager workers. The bees were collected from three non-parental colonies and kept under laboratory conditions. The bees were orally exposed continuously for five days to pyraclostrobin in syrup at concentrations of 0.125 ng a.i./μL (FG1) and 0.005 ng a.i./μL (FG2). The control bees (CTL) were fed a no-fungicide sucrose solution, and the acetone solvent control bees (CAC) received a sucrose solution containing acetone. At the end of the exposure, the midguts were sampled, fixed in Karnovsky solution, and routinely processed for transmission electron microscopy. Ultrastructural analysis demonstrated that both the fungicide concentrations altered the midgut, such as cytoplasmic vacuolization (more intense in FG1), the presence of an atypical nuclear morphology, and slightly dilated mitochondrial cristae in the bees from the FG1 and FG2 groups (both more intense in FG1). Additionally, there was an alteration in the ultrastructure of the spherocrystals (FG1), which could be the result of cellular metabolism impairment and the excretion of toxic metabolites in the digestive cells as a response to fungicide exposure. The results indicate that ingested pyraclostrobin induced cytotoxic effects in the midgut of native stingless bees. These cellular ultrastructural responses of the midgut are a prelude to a reduced survival rate, as observed in previous studies.

Keywords: digestive tract; Meliponini; mitochondria; morphology; strobilurin; sublethal effects

1. Introduction

Stingless bees are a large and diverse bee group belonging to the Meliponini tribe, of which around 550 species and 58 genera have been described worldwide [1,2]. As well as honey bees (*Apis mellifera*, Linnaeus, 1758), stingless bees are a member of the Apidae family and are the largest group of eusocial bees [3]. The geographical distribution of these bees is predominantly in tropical and subtropical regions, and they can be found in Africa [4], America [5], the Indo-Malayan region, and Australasia [2]. The distribution of stingless bees in these regions is closely related to the diversity of available flora (preferred plant families) [6].

In Brazil, there is a high diversity of about 250 described stingless bee species belonging to 29 genera, of which 20% are endemic [5]. Stingless bees are known for some remarkable characteristics, such as the incapacity to sting with a vestigial sting, well-developed defense strategies, morphological diversity, a variety of nest architectures, and importance for humans and the environment [2,7].

These native bee species are essential to conserving native flora [6,8–10] and are important for several crops [11–14]. Nowadays, the breeding and care of stingless bees (Meliponiculture) have become popular [15], which consequently has increased the commercialization and research of stingless bee products like honey, pollen, and propolis [16]. Additionally, many bioactive compounds, such as honey, propolis and geopropolis, with therapeutic properties that can contribute to treating human diseases, such as anti-inflammatory, antioxidant, and antimicrobial effects, have been discovered [17–20].

In the same way as previously mentioned, *Melipona scutellaris* (Latreille, 1811; common name—Northeast Uruçu) is an important bee species native to northeast Brazil [21], mainly due to pollination services [8] and honey production since its singular aroma and flavor are highly appreciated. The antibacterial properties of honey [22], and the antiproliferative constituents of geopropolis [23], have been described. Although *M. scutellaris* is an essential and relevant species, it is listed in the *Brazil Red Book of Threatened Species of Fauna* [24]. According to Toledo-Hernández et al. [25], many factors threaten stingless bees, highlighting pesticides (insecticides, herbicides, fungicides, biopesticides, and fertilizers), transgenic crops, deforestation, diseases and pests, competition for food resources, and climate change. This is worrying since stingless bees are not as well studied as honey bees [26].

The increasing use of fungicides, worldwide, for many years [27], means bees are more exposed to fungicides by aerial spraying or ingesting floral resources containing their residues [28]. Raimets et al. [29] detected several fungicides in beekeeping matrices. In that regard, pyraclostrobin is one of the most relevant strobilurin fungicides used in some crops that stingless bees visit (coffee, eucalyptus, and pepper) and has also been detected in pollen [30], nectar [31], and beebread [32]. Due to the inhibition of mitochondrial respiration [33], the fungicide pyraclostrobin can affect essential functions of bee physiology. A few studies have shown the side effects of fungicides on non-target stingless bees [34–38], but there is still a considerable gap between insecticide and fungicide studies [25].

Additionally, there is a lack of knowledge about the sublethal effects of fungicides on the digestive tracts of bees at a cellular level, which is responsible for vital nutritional functions [39,40] and is an entrance site for many compounds such as nutrients or toxic compounds [41]. The digestive tract of bees is divided into three compartments, the foregut or stomodaeum, the midgut or mesenteron, and the hindgut or proctodaeum [42]. Due to the importance of food digestion and the absorption of nutrients [43], the midgut can be used as a key organ for cell biomarkers evaluation in ecotoxicological studies of the sublethal effects of pesticide exposure [44–48].

According to Lourencetti et al. [49], three stingless bee species showed greater sensitivity to pesticide exposure (the insecticide neonicotinoid) than the model organism Africanized *A. mellifera* used in Brazil. Thus, stingless bees should be included in toxicological evaluation programs, as Africanized honey bees do not represent the country's vast diversity of bee species [50]. Based on this, the effects of fungicides on these native bees need to be clarified. Therefore, this study used an ultrastructural approach to evaluate the morphological alterations on a subcellular level induced by the fungicide pyraclostrobin and indications of cytotoxicity on the midgut cells after the oral exposure of *M. scutellaris* to this fungicide.

2. Materials and Methods

2.1. Meliponary

The bees used in this work were obtained from a meliponary located at the Federal University of São Carlos (UFSCar), Sorocaba campus (23°34′52.1″ S 47°31′34.7″ W), Sorocaba, Brazil. Before starting the experiments, the colonies were visually inspected to ensure a healthy status. Only colonies with similar strengths and populations that did not swarm were selected (n = 3). No chemical treatment was applied to manage the colonies, and they were kept in an urban area without pesticide application.

2.2. Stingless Bee Sampling

Forager bees of *M. scutellaris* were sampled at the entrance of nests of three non-parental colonies from the meliponary (Section 2.1) when they returned from foraging activity (Figure 1). The collections were carried out with plastic bee cages (9 × 7 cm, 250 mL) containing 120 aeration holes (3 mm) and feeders (microtube 2 µL) filled with syrup (crystal sugar 1:1 water, *w:w*) on primarily sunny days at 7:30–9:00 a.m., with temperatures with a range of 15–25 °C, throughout the summer in the Southern Hemisphere in 2021. After sampling, the cages were covered with fabric to avoid stress and transferred to the "Laboratório de Ecotoxicologia e Análise de Integridade Ambiental (LEIA)" at UFSCar and placed in an incubator at a constant temperature of 28 °C (±1) and 65% (±5) relative humidity in darkness before starting the bioassay.

Figure 1. A representative picture of the study model, the *M. scutellaris* stingless bee.

2.3. Fungicide Pyraclostrobin

The standard analytical chemical was purchased from Sigma-Aldrich (CAS Number 175013-18-0, 99.9% purity). The stock solution (1000 ng a.i./mL) was prepared using acetone as a solvent and autoclaved distilled water in proportions of 60–40%, respectively. Dilutions were performed to obtain the working solutions (0.125 ng a.i./µL and 0.005 ng a.i./µL) based on Domingues et al. [35,48]. These concentrations have been found in resources collected by bees [30,32].

2.4. Oral Exposure to Pyraclostrobin

For the bioassays, after bee sampling (Section 2.2), the feeders were removed from the cages two hours before starting oral exposure with two concentrations of the fungicide pyraclostrobin in syrup (0.125 ng a.i./µL—FG1; 0.005 ng a.i./µL—FG2). Then, the bees were randomly divided into fungicide-treated groups (FG1 and FG2), untreated controls (CTL), and solvent controls (CAC), with four replicates (n = 20 bees per cage); each experimental group contained 80 bees. The bees from CTL were fed syrup only, and the bees from CAC received syrup containing acetone (1% of the final volume) based on the recommendation of the OECD in 2013 [51]. Oral exposure was performed *ad libitum* over five days based on previous studies [35,48].

2.5. Midgut Processing for Morphological Analysis

In order to evaluate the effects of oral exposure to pyraclostrobin on *M. scutellaris*, the bees were randomly selected from all the groups (n = 6) and rendered motionless using a low temperature (4 °C) for one minute, and then dissected under a stereomicroscope at room temperature 25 °C (±1). The midguts were collected and processed for ultrastructural analysis. Light microscopy analysis was also performed according to the methodology described in the Supplementary Materials (SM). The purpose was to identify which midgut regions were more suitable for ultrastructure analysis by electron microscopy (Figure S1).

Transmission Electron Microscopy (TEM)

Three midguts for each experimental group were sampled, and the median region of each individual was subdivided into three portions (median subregions), fixed in a Karnovsky solution (2.5% glutaraldehyde—4% formaldehyde) in 0.1 M phosphate-buffered saline (pH 7.3) for 24 h at room temperature, and postfixed in 1% osmium tetroxide using the same buffer. Then, the midguts were washed in phosphate-buffered saline, dehydrated in a graded acetone series (50%, 75%, 90%, 95%, and 100%), and embedded in ultrapure resin (Araldite®). This process resulted in nine blocks per experimental group. The ultra-sections (90–60 nm) obtained from all samples were contrasted with 0.5% uranyl acetate for 20 min and lead citrate for 10 min (room temperature), and then visualized and photographed using a transmission electron microscope (Tecnai Spirit—FEI Company). Ninety regions were examined per experimental group. All steps were conducted as established in the protocol used at the Electron Microscopy Center of the Bioscience Institute (UNESP, Botucatu—Brazil), where these steps were performed.

3. Results

The TEM analysis pattern performed on the midgut epitheliums of the forager workers of *M. scutellaris* from all the experimental groups is highlighted in Figure 2. Based on the analysis, the midgut epitheliums of the bees from the CTL and CAC groups were determined to be similar. The digestive cells in both exhibited well-developed microvilli containing mitochondria with a high electron density and regular morphology, as well as mitochondria in varied formats, spherocrystals, and myelin figures in average amounts for the forager's life stage (Figure 2A,B). In contrast, the bees in the FG1 group showed large, homogeneous, electron-lucent regions in the cytoplasm of the midgut digestive cells, like cytoplasm vacuolization, while the bees in the FG2 group exhibited only small extensions of electron-lucent material in the cytoplasm (Figure 2C,D). These changes were not observed in the bees from the control groups. Additionally, digestive cells of the bees' midguts from the FG1 group showed an altered nuclei morphology with irregular shapes and several spherocrystals, which were absent in the CTL and CAC groups (Figure 2C). The midgut digestive cells from the FG2 group exhibited autophagic vacuoles in the cytoplasm (Figure 2D).

Figure 3 summarizes the apical region of the digestive cells in the midgut region, highlighting the changes mentioned above (Figure 2). The midgut digestive cells in the bees from the CTL group demonstrated standard organelle morphology, well-organized microvilli, followed by a cytoplasm rich in mitochondria, and displayed some typical spherocrystals (Figure 3A,D). Similar to those found in bees from the CTL group, the bees in FG1 and FG2 revealed digestive cells with well-organized microvilli and a cytoplasm rich in mitochondria (Figure 3B,C,E,F). However, the presence of agglomerations of spherocrystals and a mischaracterized nuclei were observed when compared to the CTL group (Figure 3B,E). Similarly, the bees from the FG2 group presented autophagic vacuoles and lipid vacuoles in the cytoplasm of digestive cells, which were not seen in the CTL group (Figure 3C,F).

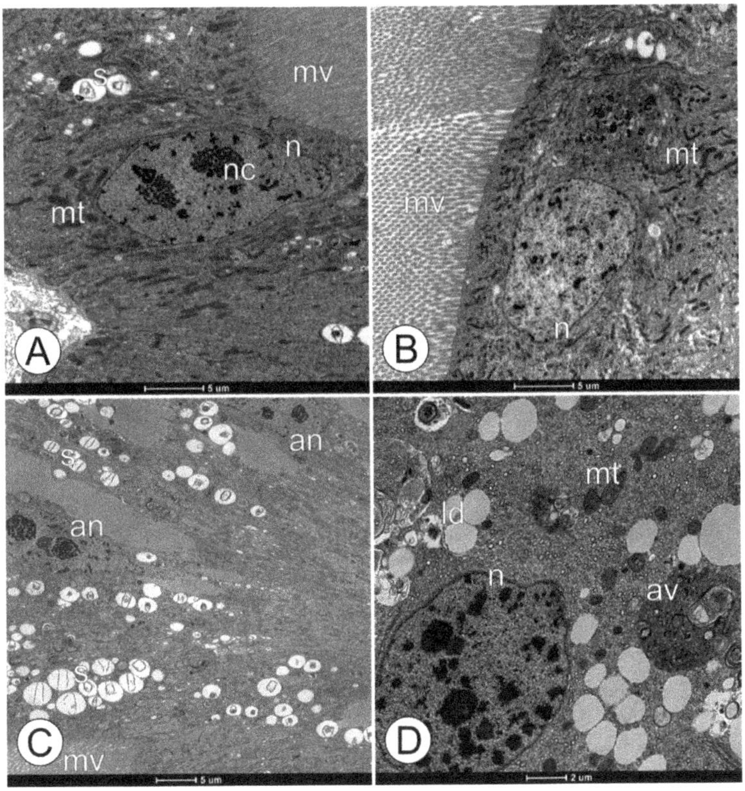

Figure 2. Ultrastructural midgut epithelium morphology in forager workers of *M. scutellaris* following five days of exposure to pyraclostrobin. (**A**) Untreated control—CTL; (**B**) solvent control—CAC; (**C**) pyraclostrobin (0.125 ng a.i./µL)—FG1; (**D**) pyraclostrobin (0.005 ng a.i./µL)—FG2. Altered nuclei (an), autophagic vacuole (av), lipid deposit (ld), microvilli (mv), mitochondria (mt), nuclei (n), nucleolus (nc), and spherocrystal (s).

Regarding the basal region of the midgut digestive cells in the midgut region, the individuals from all the experimental groups did not show any morphological changes among themselves (Figure 4). The cells in this region were characterized by the presence of evident agranular endoplasmic reticulums and mitochondria associated with the membrane forming the basal labyrinth (Figure 4A–C). In the cellular medial region, the digestive cells of the bees not exposed to the fungicide pyraclostrobin (CTL and CAC) showed a large quantity of vesiculated Golgi apparatus (Figure 4D). However, the bees from the FG1 group exhibited a more significant extension of myelin figures and spherocrystals compared to those of the CTL and FG2 groups in the medial region of the digestive cells (Figure 4E). Regarding the FG2 group, the bees demonstrated the presence of autophagic vacuoles in the digestive cells (Figure 4F).

Based on the TEM analysis of the digestive cells, the mitochondria exhibited similar ultrastructural morphologies among the groups (Figure 5). However, the mitochondrial cristae appeared tubular and slightly dilated in the bees from the FG1 and FG2 groups, becoming more evident (Figure 5A–C). The bees in the CTL and CAC groups did not exhibit this trait and were similar. Figure 6 highlights the spherocrystals present in the FG1 and FG2 bee groups; some of them appeared unstructured, and this feature was present only in the fungicide-exposed groups.

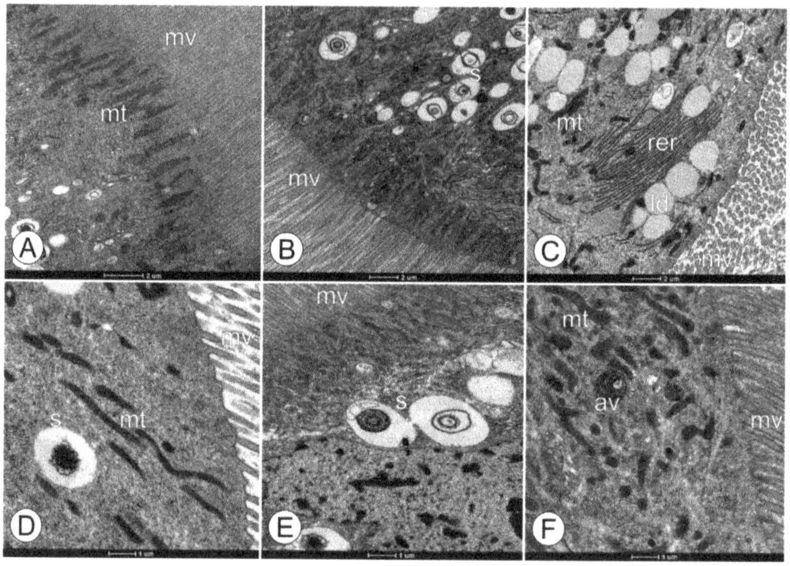

Figure 3. Ultrastructural changes in the apical region of midgut epithelium in forager workers of *M. scutellaris* following five days of exposure to pyraclostrobin. (**A**,**D**) Untreated control—CTL; (**B**,**E**) pyraclostrobin (0.125 ng a.i./μL)—FG1; (**C**,**F**) pyraclostrobin (0.005 ng a.i./μL)—FG2. Autophagic vacuole (av), lipid deposit (ld), microvilli (mv), mitochondria (mt), rough endoplasmic reticulum (rer), and spherocrystal (s).

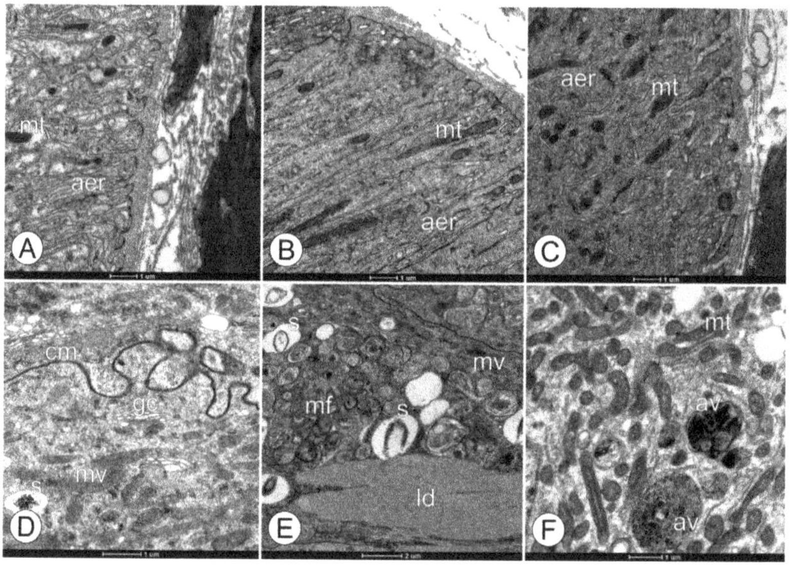

Figure 4. Ultrastructural changes in the basal and medial regions of midgut epithelium in forager workers of *M. scutellaris* following five days of exposure to pyraclostrobin. (**A**,**D**) Untreated control—CTL; (**B**,**E**) pyraclostrobin (0.125 ng a.i./μL)—FG1; (**C**,**F**) pyraclostrobin (0.005 ng a.i./μL)—FG2. Agranular endoplasmic reticulum (aer), autophagic vacuole (av), cytomembrane (cm), Golgi complexes (gc), lipid deposit (ld), mitochondria (mt), myelin figures (mf), and spherocrystal (s).

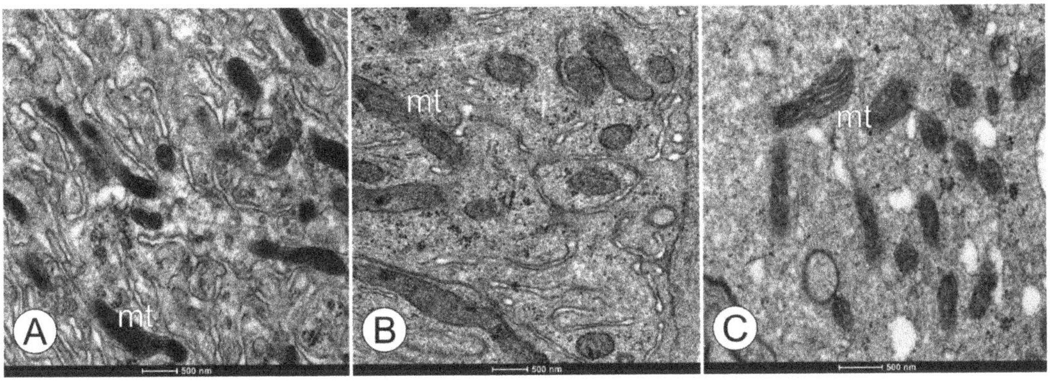

Figure 5. Ultrastructural morphology in the mitochondria of the digestive cells in the midgut epithelium in forager workers of *M. scutellaris* following five days of exposure to pyraclostrobin. (**A**) Untreated control—CTL; (**B**) pyraclostrobin (0.125 ng a.i./μL)—FG1; (**C**) pyraclostrobin (0.005 ng a.i./μL)—FG2. Mitochondria (mt).

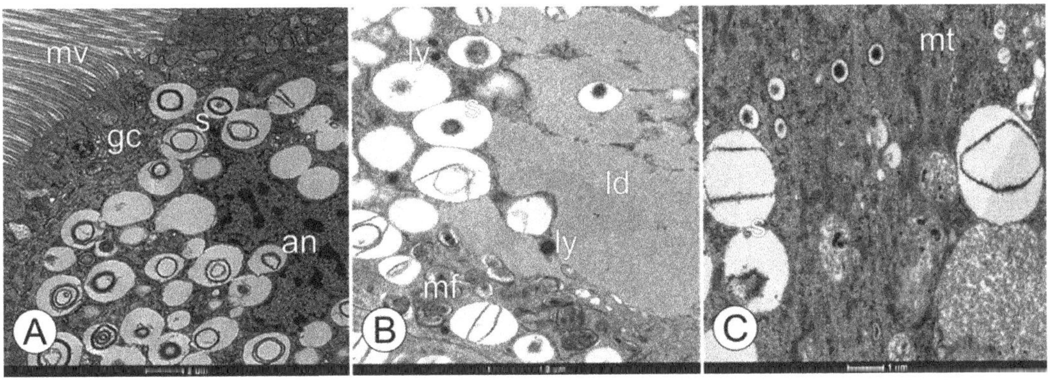

Figure 6. Ultrastructural changes in the spherocrystals of the digestive cells in the midgut epithelium in forager workers of *M. scutellaris* following five days of exposure to pyraclostrobin. (**A**,**B**) pyraclostrobin (0.125 ng a.i./μL)—FG1; (**C**) pyraclostrobin (0.005 ng a.i./μL)—FG2. Altered nuclei (an), Golgi complexes (gc), lipid deposit (ld), lysosomes (ly), microvilli (mv), mitochondria (mt), myelin figures (mf), and spherocrystals (s).

4. Discussion

Pesticides have been recognized as significant stressors, contributing to honey bee colony losses [52,53]. In Brazil, the weakness and losses in colonies of Africanized *A. mellifera* are highly associated with pesticide use [54,55]. In the same way, populations of stingless bees are also at risk due to pesticide exposure [25]. According to Rondeau and Raine [56], there are significantly more knowledge gaps regarding the risk of fungicides on some bees, particularly wild bees, compared to honey bees. From this point of view, the findings of pyraclostrobin's harmful effects on the midgut's ultrastructure are crucial to clarify some of these gaps in native stingless bees.

The electron-lucent areas, similar to cytoplasmic vacuolization, observed in the digestive cells of *M. scutellaris* indicate a cytotoxic effect caused by exposure to the higher residual concentration of pyraclostrobin continuously ingested by these bees. Similar effects were reported in *A. mellifera* workers exposed to the fungicide iprodione (dicarboximide) at a concentration of 2 mg/kg [46] and the fungicide azoxystrobin (strobilurin) at 100 μg a.i./bee [57]. Additionally, Batista et al. [58] highlighted cellular vacuolization at the zone of differentiation above the regenerative cells in the midgut of honey bees after exposure to fungicide picoxystrobin (strobilurin) at a concentration of 0.018 ng a.i./μL.

Ultrastructural analyses revealed the presence of atypical nuclear morphology, indicating a prelude to cellular death, which was more pronounced in the bees from the FG1 group. Cytotoxic effects, such as pyknotic nuclei, were also observed in the midgut epitheliums of the adult honey bees after four days of larval exposure to the fungicide pyraclostrobin at a concentration of 4.93 ng/mL, as reported by Tadei et al. [59]. The described characteristics are associated with programmed cell death (apoptosis) [60]. These responses can act as a defense mechanism against pyraclostrobin exposure, as observed with the other stressors [44,61–63]. However, even with the characteristics indicative of cytotoxicity, the cells maintained intact and probably active organelles (normal morphology), such as Golgi complexes, rough endoplasmic reticulums, and mitochondria. If the oral exposure has been prolonged, the cytotoxic effects might have expanded, and these cells would likely die.

Although the mode of action of pyraclostrobin and the other strobilurin fungicides is known to involve the inhibition of the respiratory chain [33], no drastic ultra-morphological alterations were observed in the mitochondria of the bees from the FG1 and FG2 groups. These bees exhibited only slightly dilated mitochondrial cristae, suggesting that energy production could be decreased with continued exposure. According to Zick et al. [64], the cristae morphology is linked to the bioenergetic state of the mitochondria, although there are still some gaps in understanding the key factors. A study conducted by Campbell et al. [65] highlighted that honey bees increased their mitochondrial oxygen consumption rates when exposed to the strobilurin fungicide Pristine® at concentrations of 5 ppm and higher. Ultrastructural alterations in the mitochondrial cristae were observed in the newly emerged workers of Africanized *A. mellifera* after exposure to thiamethoxam (0.001 ng/μL) during the larval phase [66], such as dilated mitochondria with a deformed shape and a loss of cristae. The mitochondrial cristae can vary from simple tubular structures to more complex lamellar structures merging with the inner boundary membrane, and their ultrastructural features have important implications for mitochondrial bioenergetics, biogenesis, and the role of mitochondria in apoptosis [67].

Myelin figures and autophagic vacuoles, or autophagosomes, are common in the digestive cells of the bee midgut, which exhibit a high turnover level of intracellular compounds, such as membranes and organelles, due to their multiple functions [68]. These cells synthesize digestive enzymes [69], compounds of the peritrophic matrix [70], and membrane protein transporters for nutrient absorption in the midgut [40]. The high absorption rate of digestive cells is linked to the longer striated border observed in this study, similar to what has been found in the midguts of other stingless bee species [71]

An alteration in the ultrastructure of the spherocrystal was observed. This alteration supports our hypothesis that fungicide exposure impairs cellular metabolism and causes the excretion of toxic products within the digestive cells. According to Serrão et al. [40], spherocrystals are relevant to maintaining osmoregulation, storing inorganic compounds, and preventing intoxication. According to the lesion index (severity and reversibility) adapted to bees [72], a score of one was assigned to spherocrystal alteration due to its association with the inactivation of toxic substances. Although in a scenario with continuous exposure, it could change and worsen.

In summary, our findings indicate that the fungicide pyraclostrobin clearly compromises the midgut of *M. scutellaris* at the ultrastructural level, offering another perspective on the effects of this fungicide on native stingless bees. These results reinforce our previous studies conducted by our research group [35,48,50,59,73]. Furthermore, our results further contribute to developing preventive safety measures to reduce the risks of pesticide use to bees. Regarding this issue, applying pesticides when bees are less active, considering weather conditions, establishing buffer zones, and implementing monitoring, integrated pest management, and habitat conservation measures, may reduce the risks of pesticides to bees.

5. Conclusions

The results of this study confirm the hypothesis that ingested pyraclostrobin induces cytotoxic effects at the ultrastructural, subcellular level of the midgut in native stingless bees. In conclusion, these cellular responses of the midgut at the tissue level may serve as a prelude to reduced bee survival rates. Therefore, it is necessary to consider the effects of fungicides on native bees and include them in protective measures to enhance regulatory decisions on risk assessment. Additionally, more studies like this would reduce the knowledge gap in how different external factors affect stingless bees.

Supplementary Materials: The following supporting information can be downloaded at: https://www.mdpi.com/article/10.3390/toxics11121028/s1. Figure S1: Midgut epithelium in forager workers of *M. scutellaris*, following five days of exposure to pyraclostrobin. (A) Untreated control—CTL; (B) solvent control—CAC; (C) pyraclostrobin 0.125 ng a.i./µL—FG1; (D) pyraclostrobin 0.005 ng a.i./µL—FG2. Asterisk (a) = apocrine secretion, black arrow = cells being released into the lumen (l), muscle (m) and villi (l). N = 6 individuals per experimental group.

Author Contributions: Conceptualization, C.E.C.D., L.V.B.I., A.G., L.S.A., O.M. and E.C.M.d.S.; data curation, C.E.C.D. and E.C.M.d.S.; formal analysis, C.E.C.D., A.G., L.S.A. and E.C.M.d.S.; funding acquisition, O.M.; investigation, C.E.C.D., L.V.B.I. and E.C.M.d.S.; methodology, C.E.C.D. and L.V.B.I.; project administration, C.E.C.D., O.M. and E.C.M.d.S.; resources, O.M. and E.C.M.d.S.; supervision, C.E.C.D., A.G., O.M. and E.C.M.d.S.; validation, C.E.C.D. and L.V.B.I.; writing—original draft, C.E.C.D., L.V.B.I., A.G., L.S.A., O.M. and E.C.M.d.S.; writing—review and editing, C.E.C.D., L.V.B.I., A.G., L.S.A., O.M. and E.C.M.d.S. All authors have read and agreed to the published version of the manuscript.

Funding: This research was funded by the São Paulo Research Foundation (FAPESP) (grant numbers 2016/15743-7 and 2017/21097-3); the Brazilian National Council for Scientific and Technological Development (CNPq) (grant number 400540/21097-3); the Slovenian Research and Innovation Agency (ARIS), research core funding no. P1-164; research for the improvement of safe food and health, the project number N4-0192; and the Public Agency for Research Activity of the Republic of Slovenia under the Contract on Stable Financing of Scientific Research Activity for the Contract Period 2022-2027, no. S-ZRD/22-27/552.

Institutional Review Board Statement: Not applicable.

Informed Consent Statement: Not applicable.

Data Availability Statement: All data associated with this research publication are available from Caio Eduardo da Costa Domingues (cecdomingues@gmail.com).

Acknowledgments: We are grateful to the "Centro de Microscopia Eletrônica—IBB/UNESP (Câmpus de Botucatu)" for all their support with transmission electron microscopy (TEM) the techniques and analysis that were carried out.

Conflicts of Interest: The authors declare no conflict of interest.

References

1. Rasmussen, C.; Cameron, S.A. Global stingless bee phylogeny supports ancient divergence, vicariance, and long distance dispersal. *Biol. J. Linn. Soc.* **2010**, *99*, 206–232. [CrossRef]
2. Grüter, C. *Stingless Bees: Their Behaviour, Ecology and Evolution*, 1st ed.; Springer: New York, NY, USA, 2020; pp. 1–385. [CrossRef]
3. Michener, C. *The Bees of the World*, 2nd ed.; Johns Hopkins University Press: Baltimore, MD, USA, 2007; pp. 1–3001.
4. Eardly, C.D. Taxonomic revision of the African stingless bee (Apoidea: Apidae: Apinae: Meliponini). *Afr. Plant Prot.* **2004**, *10*, 63–96.
5. Pedro, S.R.M. The stingless bee fauna in Brazil (Hymenoptera: Apidae). *Sociobiology* **2014**, *61*, 348–354. [CrossRef]
6. Bueno, F.G.B.; Kendall, L.; Alves, D.A.; Tamara, M.L.; Heard, T.; Latty, T.; Gloag, R. Stingless bee floral visitation in the global tropics and subtropics. *Glob. Ecol. Conserv.* **2023**, *43*, e02454. [CrossRef]
7. Roubik, D.W. Stingless bee nesting biology. *Apidologie* **2006**, *37*, 124–143. [CrossRef]
8. Imperatriz-Fonseca, V.; Saraiva, A.M.; Gonçalves, L. The Brazilian pollinators initiative and the advances for the comprehension of the role of pollinators as ecosystem services providers. *Biosci. J.* **2007**, *23*, 100–106.

9. Imperatriz-Fonseca, V.L.; Alves-dos-Santos, I.; Santos-Filho, P.S.; Engels, W.; Ramalho, M.; Wilms, W.; Aguilar, J.B.V.; Pinheiro-Machado, C.A.; Alves, D.A.; Kleinert, A.M.P. Checklist of bees and honey plants from São Paulo State, Brazil. *Biota Neotrop.* **2011**, *11*, 631–655. [CrossRef]
10. Bruckman, D.; Campbell, D.R. Floral neighborhood influences pollinator assemblages and effective pollination in a native plant. *Oecologia* **2014**, *176*, 465–476. [CrossRef]
11. Heard, T.A. The role of stingless bees in crop pollination. *Annu. Rev. Entomol.* **1999**, *44*, 183–206. [CrossRef]
12. Slaa, E.J.; Sánchez Chaves, L.A.; Malagodi-Braga, K.S.; Hofstede, F.E. Stingless bees in applied pollination: Practice and perspectives. *Apidologie* **2006**, *37*, 29–315. [CrossRef]
13. Viana, B.F.; Coutinho, J.G.E.; Garibaldi, L.A.; Castagnino, G.L.B.; Gramacho, K.P.; Silva, F.O. Stingless bees further improve apple pollination and production. *J. Pollinat. Ecol.* **2014**, *14*, 261–269. [CrossRef]
14. Giannini, T.C.; Alves, D.A.; Alves, R.; Cordeiro, G.D.; Campbell, A.J.; Awade, M.; Bento, J.M.S.; Saraiva, A.M.; Imperatriz-Fonseca, V.L. Unveiling the contribution of bee pollinators to Brazilian crops with implications for bee management. *Apidologie* **2020**, *51*, 406–421. [CrossRef]
15. Zulhendri, F.; Perera, C.O.; Chandrasekaran, K.; Ghosh, A.; Tandean, S.; Abdulah, R.; Herman, H.; Lesmana, R. Propolis of stingless bees for the development of novel functional food and nutraceutical ingredients: A systematic scoping review of the experimental evidence. *J. Funct. Foods.* **2022**, *88*, 104902. [CrossRef]
16. Rozman, A.S.; Hashim, N.; Maringgal, B.; Abdan, K. A comprehensive review of stingless bee products: Phytochemical composition and beneficial properties of honey, propolis, and pollen. *Appl. Sci.* **2022**, *12*, 6370. [CrossRef]
17. Choudhari, M.K.; Punekar, S.A.; Ranade, R.V.; Paknikar, K.M. Antimicrobial activity of stingless bee (*Trigona* sp.) propolis used in the folk medicine of Western Maharashtra, India. *J. Ethnopharmacol.* **2012**, *141*, 363–367. [CrossRef] [PubMed]
18. Rao, P.V.; Krishnan, K.T.; Salleh, N.; Gan, S.H. Biological and therapeutic effects of honey produced by honey bees and stingless bees: A comparative review. *Rev. Bras. Farmacogn.* **2016**, *26*, 657–664. [CrossRef]
19. Lavinas, F.C.; Macedo, E.H.B.C.; Sá, G.B.L.; Amaral, A.C.F.; Silva, J.R.A.; Azevedo, M.M.B.; Vieira, B.A.; Domingos, T.F.S.; Vermelho, A.B.; Carneiro, C.S.; et al. Brazilian stingless bee propolis and geopropolis: Promising sources of biologically active compounds. *Rev. Bras. Farmacogn.* **2019**, *29*, 389–399. [CrossRef]
20. Al-Hatamleh, M.A.I.; Boer, J.C.; Wilson, K.L.; Plebanski, M.; Mohamud, R.; Mustafa, M.Z. Antioxidant-based medicinal properties of stingless bee products: Recent progress and future directions. *Biomolecules* **2020**, *10*, 923. [CrossRef]
21. Alves, R.M.O.; Carvalho, C.A.L.; Souza, B.A.; Santos, W.S. Areas of natural occurrence of *Melipona scutellaris* Latreille, 1811 (Hymenoptera: Apidae) in the state of Bahia, Brazil. *An. Acad. Bras. Ciênc.* **2012**, *84*, 679–688. [CrossRef]
22. Medeiros, V.F.L.P.; Azevedo, I.M.; Rêgo, A.C.M.; Egito, E.S.T.; Araújo-Filho, I.; Medeiros, A.C. Antibacterial properties and healing effects of *Melipona scutellaris* honey in MRSA-infected wounds of rats. *Acta Cir. Bras.* **2016**, *31*, 327–332. [CrossRef]
23. Cunha, M.G.; Rosalen, P.L.; Franchin, M.; Alencar, S.M.; Ikegaki, M.; Ransom, T.; Beutler, J.A. Antiproliferative constituents of geopropolis from the bee *Melipona scutellaris*. *Planta Med.* **2016**, *82*, 190–194. [CrossRef]
24. ICMBio (Instituto Chico Mendes de Conservação da Biodiversidade). *Livro Vermelho da Fauna Brasileira Ameaçada de Extinção*; Instituto Chico Mendes de Conservação da Biodiversidade: João Pessoa, Brazil, 2018.
25. Toledo-Hernández, E.; Peña-Chora, G.; Hernández-Velázquez, V.M.; Lormendez, C.C.; Toribio-Jiménez, J.; Romero-Ramírez, Y.; León-Rodríguez, R. The stingless bees (Hymenoptera: Apidae: Meliponini): A review of the current threats to their survival. *Apidologie* **2022**, *53*, 8. [CrossRef]
26. Arena, M.; Sgolastra, F. A meta-analysis comparing the sensitivity of bees to pesticides. *Ecotoxicology* **2014**, *23*, 324–334. [CrossRef]
27. Gikas, G.D.; Parlakidis, P.; Mavropoulos, T.; Vryzas, Z. Particularities of fungicides and factors affecting their fate and removal efficacy: A review. *Sustainability* **2022**, *14*, 4056. [CrossRef]
28. Sharma, A.; Kumar, V.; Shahzad, B.; Tanveer, M.; Sidhu, G.P.S.; Handa, N.; Kohli, S.K.; Yadav, P.; Bali, A.S.; Parihar, R.D.; et al. Worldwide pesticide usage and its impacts on ecosystem. *SN Appl. Sci.* **2019**, *1*, 1446. [CrossRef]
29. Raimets, R.; Bontšutšnaja, A.; Bartkevics, V.; Pugajeva, I.; Kaart, T.; Puusepp, L.; Pihlik, P.; Keres, I.; Viinalass, H.; Mänd, M.; et al. Pesticide residues in beehive matrices are dependent on collection time and matrix type but independent of proportion of foraged oilseed rape and agricultural land in foraging territory. *Chemosphere* **2020**, *238*, 124555. [CrossRef] [PubMed]
30. Pettis, J.S.; Lichtenberg, E.M.; Andree, M.; Stitzinger, J.; Rose, R.; van Engelsdorp, D. Crop pollination exposes honey bees to pesticides which alters their susceptibility to the gut pathogen *Nosema ceranae*. *PLoS ONE* **2013**, *8*, e70182. [CrossRef]
31. Zioga, E.; Kelly, R.; White, B.; Stout, J.C. Plant protection product residues in plant pollen and nectar: A review of current knowledge. *Environ. Res.* **2020**, *189*, 109873. [CrossRef]
32. Yoder, J.A.; Jajack, A.J.; Rosselot, A.E.; Smith, T.J.; Yerke, M.C.; Sammataro, D. Fungicide contamination reduces beneficial fungi in bee bread based on an area-wide field study in honey bee, *Apis mellifera*, colonies. *J. Toxicol. Environ. Health Part A* **2013**, *76*, 587–600. [CrossRef]
33. Bartlett, D.W.; Clough, J.M.; Godwin, J.R.; Hall, A.A.; Hamer, M.; Parr-Dobrzanski, B. The strobilurin fungicides. *Pest. Manag. Sci.* **2002**, *58*, 649–662. [CrossRef]
34. Tomé, H.V.V.; Ramos, G.S.; Araujo, M.F.; Santana, W.C.; Santos, G.R.; Guedes, R.N.; Maciel, C.D.; Newland, P.L.; Oliveira, E.E. Agrochemical synergism imposes higher risk to neotropical bees than to honeybees. *R. Soc. Open Sci.* **2017**, *4*, 160866. [CrossRef]

35. Domingues, C.E.C.; Inoue, L.V.B.; Silva-Zacarin, E.C.M.; Malaspina, O. Fungicide pyraclostrobin affects midgut morphophysiology and reduces survival of Brazilian native stingless bee *Melipona scutellaris*. *Ecotoxicol. Environ. Saf.* **2020**, *206*, 111395. [CrossRef] [PubMed]
36. Prado, F.S.R.; Santos, D.M.; Almeida Oliveira, T.M.; Burgarelli, J.A.M.; Castele, J.B.; Vieira, E.M. Determination and uptake of abamectin and difenoconazole in the stingless bee *Melipona scutellaris* Latreille (1811) via oral and topic acute exposure. *Environ. Pollut.* **2020**, *265*, 114313. [CrossRef] [PubMed]
37. Almeida, C.H.S.; Haddi, K.; Toledo, P.F.S.; Rezende, S.M.; Santana, W.C.; Guedes, R.N.C.; Newland, P.L.; Oliveira, E.E. Sublethal agrochemical exposures can alter honey bees' and Neotropical stingless bees' color preferences, respiration rates, and locomotory responses. *Sci. Total Environ.* **2021**, *779*, 146432. [CrossRef]
38. Brigante, J.; Costa, J.O.; Espíndola, E.L.G.; Daam, M.A. Acute toxicity of the insecticide abamectin and the fungicide difenoconazole (individually and in mixture) to the tropical stingless bee *Melipona scutellaris*. *Ecotoxicology* **2021**, *30*, 1872–1879. [CrossRef] [PubMed]
39. Serrão, J.E.; Cruz-Landim, C. Ultrastructure of digestive cells in stingless bees of various ages (Hymenoptera, Apidae, Meliponinae). *Cytobios* **1996**, *88*, 161–171.
40. Serrão, J.E.; Ronnau, M.; Neves, C.A.; Campos, L.A.; Zanuncio, J.C. Ultrastructure of anterior midgut region of corbiculate bees (Hymenoptera: Apidae). *Ann. Entomol. Soc. Am.* **2008**, *101*, 915–921. [CrossRef]
41. Denecke, S.; Swevers, L.; Douris, V.; Vontas, J. How do oral insecticidal compounds cross the insect midgut epithelium? *Insect Biochem. Mol. Biol.* **2018**, *103*, 22–35. [CrossRef] [PubMed]
42. Cruz-Landim, C. *Abelhas: Morfologia e Função de Sistemas*, 1st ed.; UNESP: São Paulo, Brazil, 2009; pp. 1–416. [CrossRef]
43. Terra, W.R.; Ferreira, C. Evolutionary trends of digestion and absorption in the major insect orders. *Arthropod Struct. Dev.* **2020**, *56*, 100931. [CrossRef] [PubMed]
44. Motta, J.V.d.O.; Carneiro, L.S.; Martínez, L.C.; Bastos, D.S.S.; Resende, M.T.C.S.; Castro, B.M.C.; Neves, M.M.; Zanuncio, J.C.; Serrão, J.E. Midgut cell damage and oxidative stress in *Partamona helleri* (Hymenoptera: Apidae) workers caused by the insecticide lambda-cyhalothrin. *Antioxidants* **2023**, *12*, 1510. [CrossRef] [PubMed]
45. Serra, R.S.; Cossolin, J.F.S.; Resende, M.T.C.S.; Castro, M.A.; Oliveira, A.H.; Martínez, L.C.; Serrão, J.E. Spiromesifen induces histopathological and cytotoxic changes in the midgut of the honeybee *Apis mellifera* (Hymenoptera: Apidae). *Chemosphere* **2021**, *270*, 129439. [CrossRef] [PubMed]
46. Carneiro, L.S.; Martínez, L.C.; Gonçalves, W.G.; Santana, L.M.; Serrão, J.E. Ecotoxicology and Environmental Safety the fungicide iprodione affects midgut cells of non-target honey bee *Apis mellifera* workers. *Ecotoxicol. Environ. Saf.* **2020**, *189*, 109991. [CrossRef] [PubMed]
47. Castro, M.B.A.; Martinez, L.C.; Cossolin, J.F.S.; Serra, R.S.; Serrão, J.E. Cytotoxic effects on the midgut, hypopharyngeal, glands and brain of *Apis mellifera* honey bee workers exposed to chronic concentrations of lambda-cyhalothrin. *Chemosphere* **2020**, *248*, 126075. [CrossRef] [PubMed]
48. Domingues, C.E.C.; Inoue, L.V.B.; Silva-Zacarin, E.C.M.; Malaspina, O. Foragers of Africanized honeybee are more sensitive to fungicide pyraclostrobin than newly emerged bees. *Environ. Pollut.* **2020**, *266*, 115267. [CrossRef] [PubMed]
49. Lourencetti, A.P.S.; Azevedo, P.; Miotelo, L.; Malaspina, O.; Nocelli, R.C.F. Surrogate species in pesticide risk assessments: Toxicological data of three stingless bees species. *Environ. Pollut.* **2023**, *318*, 120842. [CrossRef]
50. Assis, J.C.; Tadei, R.; Menezes-Oliveira, V.B.; Silva-Zacarin, E.C.M. Are native bees in Brazil at risk from the exposure to the neonicotinoid imidacloprid? *Environ. Res.* **2022**, *212*, 113127. [CrossRef]
51. OECD (Organization for Economic Co-operation and Development). Test No. 213: Honeybees, Acute Oral Toxicity Test. In *OECD Guidelines for the Testing of Chemicals*; OECD Publishing: Paris, France, 1998.
52. Sánchez-Bayo, F.; Goulson, D.; Pennacchio, F.; Nazzi, F.; Goka, K.; Desneux, N. Are bee diseases linked to pesticides?—A brief review. *Environ. Int.* **2016**, *89–90*, 7–11. [CrossRef]
53. Insolia, L.; Molinari, R.; Rogers, S.R.; Williams, G.R.; Chiaromonte, F.; Calovi, M. Honey bee colony loss linked to parasites, pesticides and extreme weather across the United States. *Sci. Rep.* **2022**, *12*, 20787. [CrossRef]
54. Maggi, M.; Antúnez, K.; Invernizzi, C.; Aldea, P.; Vargas, M.; Negri, P.; Brasesco, C.; De Jong, D.; Message, D.; Teixeira, E.W.; et al. Honeybee health in South America. *Apidologie* **2016**, *47*, 835–854. [CrossRef]
55. Castilhos, D.; Bergamo, G.C.; Kastelic, J.P. Honey bee colony losses in Brazil in 2018-2019. *Braz. J. Anim. Environ. Res.* **2021**, *4*, 5017–5041. [CrossRef]
56. Rondeau, S.; Raine, N.E. Fungicides and bees: A review of exposure and risk. *Environ. Int.* **2022**, *165*, 107311. [CrossRef]
57. Serra, R.S.; Martínez, L.C.; Cossolin, J.F.S.; Resende, M.T.C.S.; Carneiro, L.S.; Fiaz, M.; Serrão, J.E. The fungicide azoxystrobin causes histopathological and cytotoxic changes in the midgut of the honey bee *Apis mellifera* (Hymenoptera: Apidae). *Ecotoxicology* **2023**, *32*, 234–242. [CrossRef]
58. Batista, A.C.; Domingues, C.E.C.; Costa, M.J.; Silva-Zacarin, E.C.M. Is a strobilurin fungicide capable of inducing histopathological effects on the midgut and Malpighian tubules of honey bees? *J. Apic. Res.* **2020**, *59*, 834–843. [CrossRef]
59. Tadei, R.; Menezes-Oliveira, V.B.; Silva-Zacarin, E.C.M. Silent effect of the fungicide pyraclostrobin on the larval exposure of the non-target organism Africanized *Apis mellifera* and its interaction with the pathogen *Nosema ceranae* in adulthood. *Environ. Pollut.* **2020**, *267*, 115622. [CrossRef]
60. Elmore, S. Apoptosis: A review of programmed cell death. *Toxicol. Pathol.* **2007**, *35*, 495–516. [CrossRef]

61. Domingues, C.E.C.; Abdalla, F.C.; Balsamo, P.J.; Pereira, B.V.R.; Hausen, M.A.; Costa, M.J.; Silva-Zacarin, E.C.M. Thiamethoxam and picoxystrobin reduce the survival and overload the hepato-nephrocitic system of the Africanized honeybee. *Chemosphere* **2017**, *186*, 994–1005. [CrossRef] [PubMed]
62. Carneiro, L.S.; Martinez, L.C.; Oliveira, A.H.; Cossolin, J.F.S.; Resende, M.T.C.S.; Gonçalves, W.G.; Medeiros-Santana, L.; Serrão, J.E. Acute oral exposure to imidacloprid induces apoptosis and autophagy in the midgut of honey bee *Apis mellifera* workers. *Sci. Total Environ.* **2022**, *818*, 152847. [CrossRef] [PubMed]
63. Gao, J.; Guo, Y.; Chen, J.; Diao, Q.-Y.; Wang, Q.; Dai, P.-L.; Zhang, L.; Li, W.-M.; Wu, Y.-Y. Acute oral toxicity, apoptosis, and immune response in nurse bees (*Apis mellifera*) induced by flupyradifurone. *Front. Physiol.* **2023**, *14*, 1150340. [CrossRef]
64. Zick, M.; Rabl, R.; Reichert, A.S. Cristae formation—Linking ultrastructure and function of mitochondria. *Biochim. Biophys. Acta, Mol. Cell Res.* **2009**, *1793*, 5–19. [CrossRef]
65. Campbell, J.B.; Nath, R.; Gadau, J.; Fox, T.; Degrandi-Hoffman, G.; Harrison, J.F. The fungicide Pristine® inhibits mitochondrial function in vitro but not flight metabolic rates in honey bees. *J. Insect Physiol.* **2016**, *86*, 11–16. [CrossRef]
66. Friol, P.S.; Catae, A.F.; Tavares, D.A.; Malaspina, O.; Roat, T.C. Can the exposure of *Apis mellifera* (Hymenoptera, Apiadae) larvae to a field concentration of thiamethoxam affect newly emerged bees? *Chemosphere* **2017**, *185*, 56–66. [CrossRef]
67. Frey, T.G.; Mannella, C.A. The internal structure of mitochondria. *Trends Biochem. Sci.* **2000**, *25*, 319–324. [CrossRef] [PubMed]
68. Oliveira, A.H.; Gonçalves, W.G.; Fernandes, K.M.; Barcellos, M.S.; Sampaio, W.M.S.; Lopes, M.P.; Martins, G.F.; Serrão, J.E. Morphology and morphometry of the midgut in the stingless bee *Friesella schrottkyi* (Hymenoptera: Apidae). *Insects* **2019**, *10*, 73. [CrossRef] [PubMed]
69. Jimenez, D.R.; Gilliam, M. Age-related changes in midgut ultrastructure and trypsin activity in the honey bee, *Apis mellifera*. *Apidologie* **1989**, *20*, 287–303. [CrossRef]
70. Teixeira, A.D.; Marques-Araujo, S.; Zanuncio, J.C.; Serrão, J.E. Peritrophic membrane origin in adult bees (Hymenoptera): Immunolocalization. *Micron* **2015**, *68*, 91–97. [CrossRef] [PubMed]
71. Serrão, J.E.; Cruz-Landim, C. The striated border of digestive cells in adult stingless bees (Hymenoptera, Apidae, Meliponinae) Cytobios 1995, 83, 229–235. *83*.
72. Grella, T.C.; Soares-Lima, H.M.; Malaspina, O.; Nocelli, R.C.F. Semi-quantitative analysis of morphological changes in bee tissues: A toxicological approach. *Chemosphere* **2019**, *236*, 124255. [CrossRef]
73. Inoue, L.V.B.; Domingues, C.E.C.; Gregorc, A.; Silva-Zacarin, E.C.M.; Malaspina, O. Harmful effects of pyraclostrobin on the fat body and pericardial cells of foragers of Africanized honey bee. *Toxics* **2022**, *10*, 530. [CrossRef]

toxics MDPI

Article

Toxic Effects of Methylene Blue on the Growth, Reproduction and Physiology of *Daphnia magna*

Shuhui Li [1,†], Yixin Cui [2,†], Min Wen [2] and Gaohua Ji [1,2,*]

1 National Demonstration Center for Experimental Fisheries Science Education, Shanghai Ocean University, Shanghai 201306, China; 1911202@st.shou.edu.cn
2 Engineering Research Center of Environmental DNA and Ecological Water Health Assessment, Shanghai Ocean University, Shanghai 201306, China; m210100130@st.shou.edu.cn (Y.C.); m200100099@st.shou.edu.cn (M.W.)
* Correspondence: ghji@shou.edu.cn
† These authors contributed equally to this work.

Abstract: Methylene blue (MB) is a disinfectant used in aquaculture to prevent and treat fish diseases. However, the release of MB can pose a risk to the receiving water bodies. Zooplankton are the most sensitive organisms among aquatic life. Hence, this study examined the acute and chronic toxic effects of MB on zooplankton using *Daphnia magna* (*D. magna*) as a test organism to provide basic data for risk assessment. The results show that 48 h-EC_{50} and 24 h-LC_{50} were 61.5 ± 2.3 and 149.0 ± 2.2 μg/L, respectively. Chronic exposure to MB affected the heart rate, beat frequency of the thoracic limbs, and reproductive ability of *D. magna* at environmental concentrations higher than 4.7 μg/L. The cumulative molts, time to production of the first brood, and total number of living offspring were affected at different MB concentrations, while "abortions" were observed in high-exposure groups. The activity of superoxide dismutase was increased, while glutathione S-transferase activity was stimulated at low concentrations and inhibited at high concentrations. In addition, the malondialdehyde content increased with increasing concentrations of MB. Our findings demonstrate the impact of MB on the reproduction and growth of freshwater species, as well as their physiological responses. These results have implications for establishing guidelines on the use of MB in aquaculture and setting discharge standards.

Keywords: antimicrobial dyes; zooplankton; life table parameters; oxidative damage

1. Introduction

Methylene blue (MB) is an aromatic dye with the molecular formula $C_{16}H_{18}ClN_3S$ and a benzene ring with redox characteristics. It is commonly used in various industries, including the textile, pharmaceutical, paper, dyeing, printing, paint, pharmaceutical, and food industries, as well as in medical research and aquaculture [1–3].

In aquaculture, MB is a popular environmental disinfectant. Disinfectants frequently used in aquaculture can be broadly categorized into three groups: halogens, dyes, and surfactants. Among the dyes, trityl methane dyes such as malachite green and crystal violet are prohibited substances due to their long residual time and toxic side effects, so MB has emerged as one of the best alternatives. It is commonly used to disinfect aquaculture environments because the ionic compound generated in the aqueous solution can compete with microbial enzyme systems for hydrogen ions and inactivate the enzymes, which results in the loss of viability of microorganisms. Furthermore, MB, as an antifungal drug, is used to prevent and control fish diseases such as saprolegniasis, ichthyophthiriasis, chilodonelliosis, and gill disease to reduce mortality in fish during transportation [4].

The environment of aquaculture water directly affects whether aquatic animals can grow quickly and healthily, and the quality of the water is closely related to the occurrence of diseases. As the water quality deteriorates, it encourages the production of different types of pathogens, endangering the growth and development of farmed animals. Disinfectants play a crucial role in aquaculture, from cleansing water bodies to bathing fish before transferring them to ponds, disease management, and even water quality regulation. Disinfectants should be used rationally and scientifically to eradicate or destroy pathogenic microorganisms in the aquaculture environment and stop the spread of disease.

Malachite green is listed as a banned drug in some countries [5] due to its potential toxicity to the water environment and human health. It belongs to the same group of tritylene-based dyes as methylene blue and has a long residual time, as well as teratogenic, carcinogenic, and mutagenic risks [6]. The widespread use of methylene blue can result in residues in water bodies that persist in the environment [7], and the associated water contamination problems cannot be ignored. Additionally, the color of malachite green can prevent sunlight from passing through the water body, resulting in reduced dissolved oxygen and inhibited photosynthesis [8]. This can lead to reduced diversity in biological communities [9–12] and interfere with the normal functioning of aquatic ecosystems.

Recent studies have reported varying toxic effects of MB on different aquatic organisms (Supplementary Materials, Table S1). Perlberg et al. [13] found that MB was teratogenic to *Pterophyllum scalare*, reporting that exposure to a 5 ppm concentration resulted in a higher incidence of non-inflatable swim bladders. In contrast, Soltanian et al. [14] found that goldfish (*Carassius auratus*) exposed to a 2 mg/L solution of methylene blue for 21 days had significantly reduced lethality caused by *Aeromonas hydrophila*. However, the fish also exhibited significantly lower levels of neutrophils and aspartate aminotransferase, indicating some immunosuppressive effects and potential harm to their health. Comparing the toxicity of MB in various aquatic animals, it is evident that although it is a valuable tool in aquaculture, it still has some negative consequences. Currently, the U.S. Food and Drug Administration (FDA), EU Directive 96/23/EC, and Japan's "positive list" have established guidelines for detecting MB residues in aquatic products [15]. Despite being repeatedly banned, MB continues to be used due to its low cost and high efficiency [16,17]. In addition, many countries, such as China, have yet to establish limits on the use of methylene blue in animal-derived food products.

Freshwater zooplankton are a critical component of aquatic environments, playing a vital role in the material cycle and energy flow. Despite their importance, there are still unanswered questions regarding the ecotoxicity of MB in this group of organisms. *Daphnia magna* is a typical representative of zooplankton [18], with a transparent body. It feeds on algae, which helps improve water quality, and is a natural bait for filter-feeding fish [19,20]. *D. magna* lays eggs in its brood chamber, where they develop until the eye point appears, forming an embryo. The embryo then develops into a neonate and is released from the parent's body (Supplementary Materials, Figure S1).

Daphnia magna has been extensively used for toxicity testing, identifying water pollution, and creating water quality standards due to its sensitivity to chemical exposure [21,22]. To assess the toxicity of chemicals to *D. magna*, a range of parameters were determined, including reproduction, physiology, swimming behavior, and biochemistry. Growth and reproductive capacity are sensitive indicators of chronic toxicity for *D. magna* [23] and are also important factors in assessing population growth capacity [24], including parameters such as lifespan, total number of living offspring, time to first brood, cumulative molts, reproductive rate, and number of aborted eggs. Swimming behavior parameters consist of swimming activity, swimming time, swimming speed, and so on. Among the physiological parameters are feeding rate, heart rate, thoracic limb activity, post-abdominal claw movement, and compound eye activity. Biochemical parameters can be categorized as enzymatic or non-enzymatic. Quantitative studies on enzyme activity variations in *D. magna* have been conducted to determine the ecological risk of chemicals [25–27].

The safety of the aquatic environment is a pressing issue in the 21st century, with significant ramifications for society and the global economy. Yee et al. [28] highlighted the persistent issue of the negative effects of man-made surface biological contamination. To address this issue, we used *D. magna* as a test organism to investigate the acute and chronic toxic effects of MB and calculated the median effect concentration (EC_{50}), median lethal concentration (LC_{50}), and no observed effective concentration (NOEC) to provide basic data for aquatic ecological risk assessment. Additionally, our findings provide a reference for the safe and rational use of MB and ecological diversity conservation.

2. Materials and Methods

2.1. Chemicals

The MB solution used in this study was purchased from Sangon Biotech Company (Shanghai, China), with a blue color and purity $\geq$ 98%. The basic physical and chemical properties of MB are presented in Table 1. A stock solution with a concentration of 100 mg/L was prepared with ultrapure water and stored at 4 °C in a refrigerator protected from light.

Table 1. Physical and chemical properties of methylene blue (MB).

Chemical Name	Molecular Formula	Molecular Weight	CAS Registry Number	Solubility (g/L)	Chemical Structure
Chloro-3,7-bis(dimethylamino) phenothiazine-5-buzz-trishydrate	$C_{16}H_{18}ClN_3S·3(H_2O)$	373.9	7220-79-3	50 (20 °C)	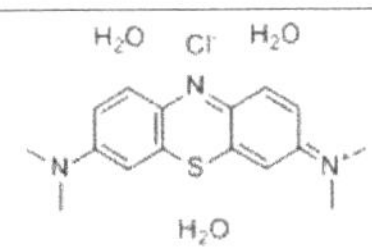

2.2. Preparation of *Daphnia magna*

D. magna were cultured in the laboratory for over 3 generations, all derived from a single parent. They were cultured in a 500 mL glass beaker placed in an incubator at 20 ± 1 °C, with a 16:8 light/dark photoperiod and a light intensity of 3000 Lux. EPA medium was used as the culture medium, and it was renewed 2–3 times a week. The animals were fed daily on *Chlorella vulgaris* (at a concentration of 2.5×10^6 cells/mL), and dead *D. magna* and impurities at the bottom were cleaned daily. Larger, more active females with more eggs were selected 24 h before the experiment and incubated in a separate beaker. The newly produced juveniles (6~24 h) were used as test organisms.

2.3. Acute Experiments

In 48 h immobilization toxicity tests, referring to the Test No. 221 guidance document [29], 6 concentration groups (30, 39.6, 52.3, 69, 91.1, and 120 μg/L) and a control group were set up according to the pilot experiment. The test system consisted of 50 mL beakers containing 30 mL of MB solution. Ten healthy neonates (<24 h) were randomly placed in each beaker without feeding, and three replicates were set up for each group. After 48 h, the morphology of *D. magna* was observed under a microscope, and the inhibition rate was calculated (inhibited being defined as having a heartbeat but not swimming). The 48 h-EC_{50} value was obtained through curve fitting.

In 24 h lethality toxicity tests, after pilot experiments, 6 treatment groups (110, 124, 140, 157, 177, and 200 μg/L) and a control group were established. The test system consisted of 50 mL beakers containing 30 mL of MB solution. Ten healthy neonates (<24 h) were randomly placed in each beaker without feeding, and three replicates were set up for each group. After 24 h, the morphology of *D. magna* was observed under a microscope, and the mortality rate was calculated (with the criterion of no heartbeat as death). The 24 h-LC_{50} value was obtained by curve fitting.

2.4. Chronic Experiments

According to the value of 24 h-LC_{50}, 6 treatment groups (1.5, 2.7, 4.7, 8.4, 15, and 26.7 μg/L) and one blank control group were set up using the equal logarithmic spacing method. The experimental system consisted of 50 mL beakers containing 30 mL of MB solution, with one healthy neonate (<24 h) randomly placed in each beaker, and 10 replicates were set up for each concentration group, for a total of 70 neonates (7 × 10) used in the experiment, which lasted 21 days. *Chlorella vulgaris* were fed daily at a density of 2.5×10^6 cells/mL. During the experiment, a semi-static exposure test was used, and the exposure solution was replaced every 2 days. The growth and reproductive status were observed and recorded daily, including cumulative molts, time of the first brood, number of offspring in the first brood, total number of broods, and number of living offspring per brood, to calculate the intrinsic growth rate (r_m) of the population. The newborns were promptly removed after birth.

At the end of the experiment, each *D. magna* was placed in the groove of a single concave glass slide, and an appropriate amount of MB exposure solution was dropped in. The body length (from the helmet to the front end of the tail spine, excluding the tail spine), the heart rate, and the frequency of thoracic limb movement were measured using a stereomicroscope and VistarImage software. Each *D. magna* was measured 3 times.

2.5. Measurement of Physiological Parameters

Antioxidant damage tests were conducted on *D. magna* in the 8.4, 15, and 26.7 μg/L exposure concentrations and the blank control group based on the results of the chronic test. For these tests, 100 healthy neonates (age 6~24 h) were randomly placed in a 250 mL beaker containing 200 mL of MB solution, without feeding, and there were 3 replicates for each group. After 24 h, the *D. magna* were collected, washed 3 times, and transferred to a 1.5 mL tube containing 0.81 mL of physiological saline. The tube was then placed at −40 °C for 12 h and then at 4 °C for 1 h to thaw. An ultrasonic cell disrupter was used to crush the tissue. The prepared 10% tissue homogenate was centrifuged at 2500 rpm for 15 min in a high-speed frozen centrifuge at 4 °C. The supernatant obtained was the crude enzyme solution for measurement. Superoxide dismutase (SOD) activity, glutathione S-transferase (GST) activity, and malondialdehyde (MDA) content were measured using test kits purchased from Nanjing Jiancheng Bioengineering Institute, Nanjing, China.

2.6. Statistical Analysis

The experimental data were analyzed and statistically processed using SPSS 23.0 and Excel software. The experimental results were subjected to linear or nonlinear fitting analysis using OriginPro 9.1 software. The most suitably fitted model was selected based on the R^2 value of the fitting curve being closer to 1 and the p value being smaller. The trend of inhibition and mortality of *D. magna* under exposure to different concentrations of methylene blue was fitted using the Hill function (Equation (1)) to obtain the 48 h-EC_{50} and 24 h-LC_{50} values:

$$f(x) = \frac{1}{1 + \left(\frac{EC50}{x}\right)^m} \quad (1)$$

where $f(x)$ refers to the mortality (or immobilization rate) of *D. magna*, x refers to the concentration of MB solution (μg/L), and m refers to the curve shape parameter.

The Log3P1 model (Equation (2)) was used to fit the trend of heart rate and thoracic limb activity in the chronic treatment:

$$y = a - b\ln(x + c) \quad (2)$$

where y refers to the heart rate or beat frequency of the thoracic limb (times/min), x refers to the concentration of MB solution (μg/L), and a, b and c are constants.

The intrinsic rate of population increase (r_m) was initially calculated by Equation (3), and then the precise value was obtained through the stepwise approximation method in Equation (4) [30,31]:

$$r_m = \frac{\ln R_0}{T}, \; R_0 = \sum_0^{\infty} m_x l_x, \; T = \frac{\sum_0^{\infty} x l_x m_x}{R_0} \quad (3)$$

$$\sum_{x=0}^{n} e^{-rm_x} l_x m_x = 1 \quad (4)$$

where x is the age of *D. magna* (d), lx is the survival rate at age x, m_x is the fecundity at age x, R_0 is the net reproduction rate, and T is the generation time (d).

The normality of the parameters was tested using the Shapiro–Wilks test. A one-way ANOVA followed by an LSD post hoc test was used to analyze the differences between the blank control and various MB concentration groups, and the experimental results were presented as mean ± standard error.

3. Results

3.1. Acute Immobilization Toxicity Tests

The immobilization rate of *D. magna* increased with increasing MB concentration (Figure 1). The 48 h-EC_{50} of MB for *D. magna* was determined to be 61.5 ± 2.3 μg/L.

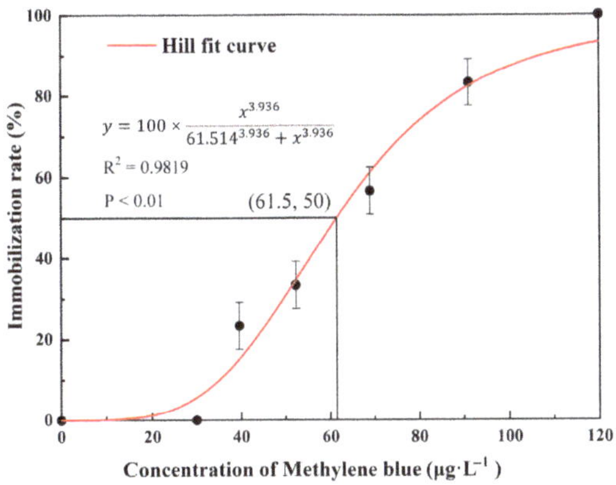

Figure 1. Trend of immobilization rate fitted by Hill function in acute toxicity tests. Means and standard errors are shown.

3.2. Acute Lethality Toxicity Tests

The mortality of *D. magna* increased with the increased MB concentration (Figure 2). The 24 h-LC_{50} of MB for *D. magna* was determined to be 149.0 ± 2.2 μg/L.

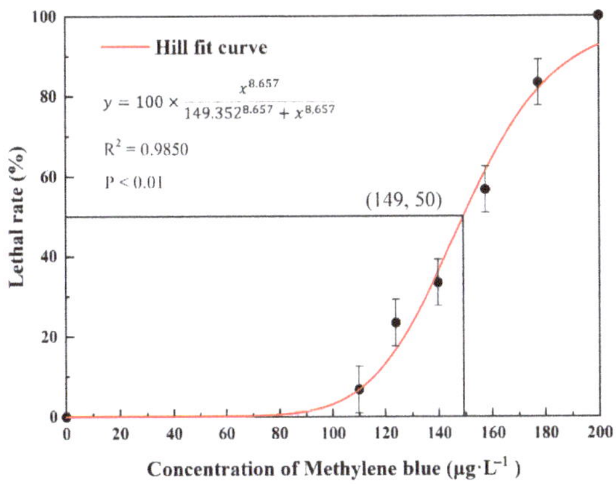

Figure 2. Trend of mortality rate fitted by Hill function in acute toxicity tests. Means and standard errors are shown.

3.3. Damage Caused by MB to D. magna Bodies in Acute Toxicity Tests

In the acute exposure test, the tested *D. magna* suffered from various degrees of damage. Some individuals had holes in their carapace and lost their thoracic limbs, while others had a swollen carapace and blue residue on their antennae, appendages, intestines, and carapace. The bodies of *D. magna* that had stopped beating were white and sank to the bottom of the exposure solution (Figure 3).

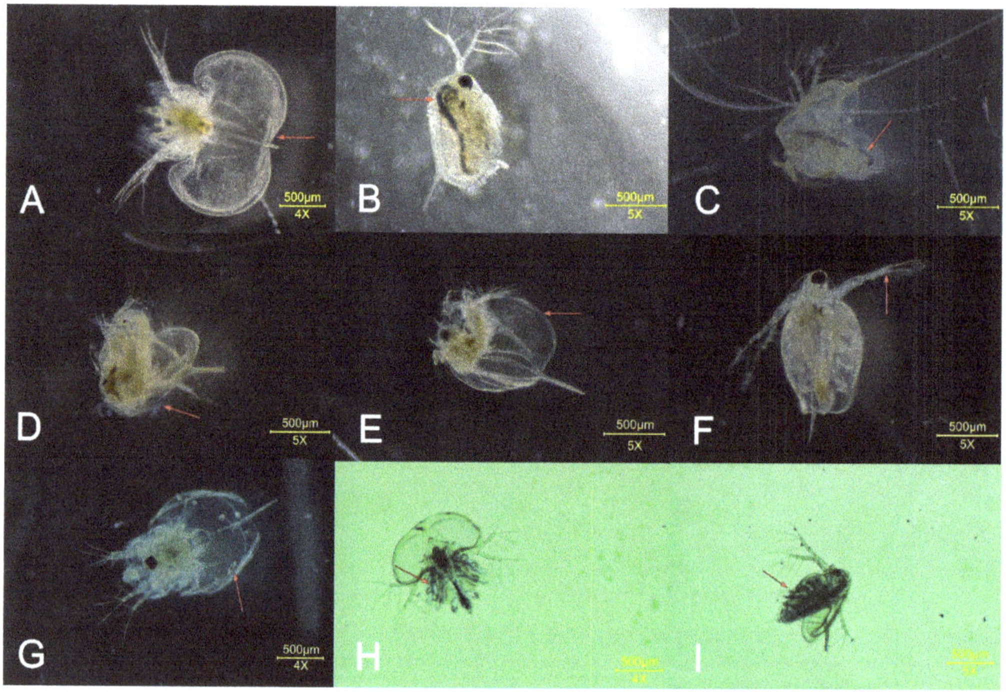

Figure 3. Damage to *D. magna* caused by MB in acute toxicity tests: (**A**) swollen and whitish carapace; (**B**) blue intestines; (**C**) swollen carapace and blue intestines; (**D**) carapace stuck with blue substance; (**E**) swollen carapace and partial loss of thoracic limb; (**F**) blue substance sticking to second antenna; (**G**) body turned blue throughout, with swollen carapace and holes; (**H**) swollen carapace and blue substance sticking to thoracic limb; (**I**) blue deposits in intestines and thoracic limbs.

3.4. Chronic Toxicity of Methylene Blue in D. magna

3.4.1. Heart Rate and Thoracic Limb Activity

The heart rate of *D. magna* decreased as the MB concentration increased (Figure 4). At 4.7 μg/L, the heart rate of the exposed group (269 ± 2.35 times/min) was significantly different ($p < 0.05$) from that of the control group (298 ± 10.15 times/min). The lowest heart rate was observed in the 26.7 μg/L MB treated group (256 ± 5.57 times/min), which was significantly different ($p < 0.05$) from the control group, as well as the 1.5, 2.7, 4.7, and 8.4 μg/L MB treated groups. Similarly, as the MB concentration increased, the thoracic limb beat frequency decreased. At 4.7 μg/L, the beat frequency of the exposed group (289 ± 10.75 times/min) was significantly different ($p < 0.05$) from that of the control group (301 ± 4.95 times/min). The lowest beat frequency was observed in the 26.7 μg/L exposed group (280.2 ± 1.92 times/min), which was significantly different ($p < 0.05$) from that of the control group as well as the 1.5, 2.7, 4.7, and 8.4 μg/L treatment groups.

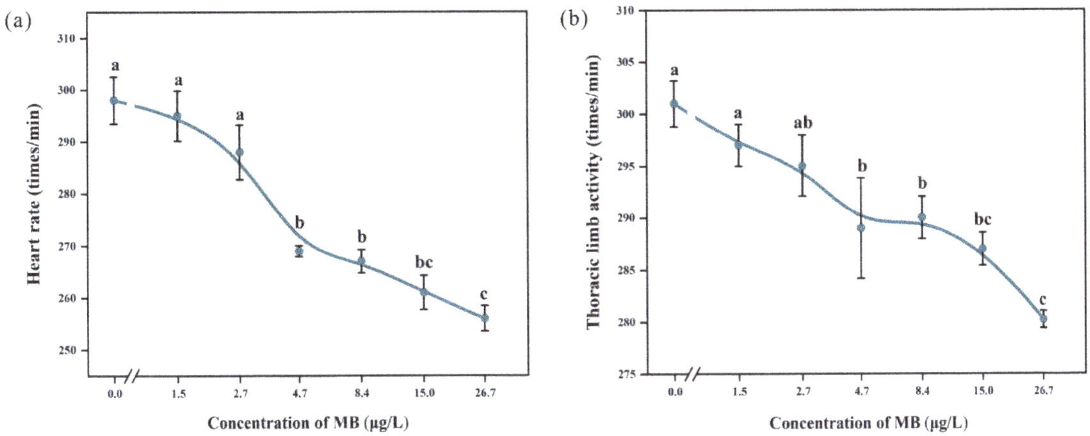

Figure 4. Effect of MB on (**a**) heart rate and (**b**) thoracic limb activity of *D. magna* (mean ± SE, $n = 10$); significant differences ($p < 0.05$) between different treatment groups and the control group are indicated by different letters.

The inhibitory effect of MB on heart rate and thoracic limb beat frequency was dose-dependent (Figure 5).

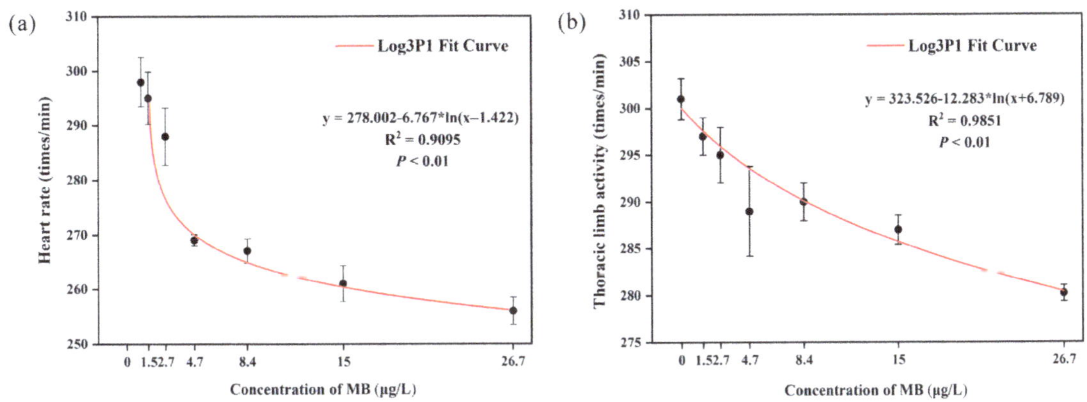

Figure 5. Effects of MB on (**a**) heart rate and (**b**) thoracic limb activity of *D. magna* after fitting with the Log3P1 curve. Means and standard errors are shown.

3.4.2. Body Length

After exposure to methylene blue for 21 days, the body length of *D. magna* in the 15 μg/L group increased (3.472 ± 0.078 mm) compared to the control group (3.462 ± 0.094 mm), but there was no significant difference ($p > 0.05$). The body length of *D. magna* in all other groups was shorter than that of the control group, and the body length in the highest concentration group (26.7 μg/L) was significantly shorter (3.206 ± 0.211 mm, $p < 0.05$; Figure 6).

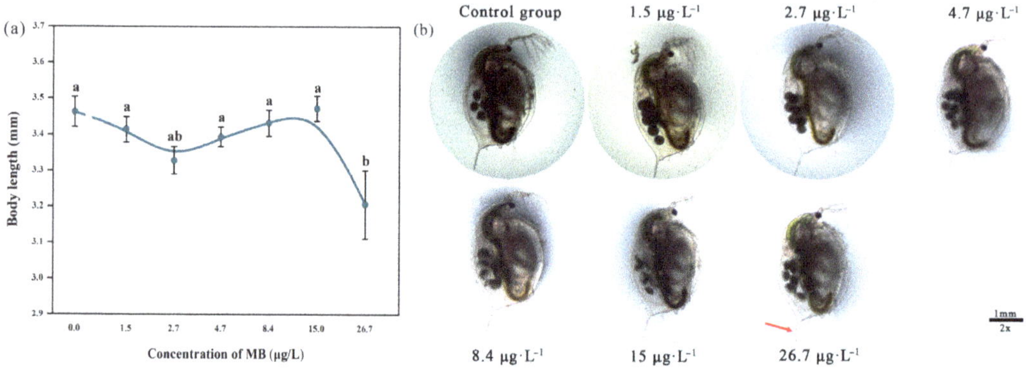

Figure 6. Effect of MB on (**a**) body length and (**b**) morphology of *D. magna* in each group at 21 days (mean ± SE, $n = 10$); significant differences ($p < 0.05$) between different treatment groups and the control group are indicated by different letters.

3.4.3. Reproduction

The total number of molts of *D. magna* showed a trend of increasing and then decreasing with increasing MB concentration. When the MB concentration was 8.4 μg/L, the number of molts reached a peak at 9.1 ± 1.52 times, but there was no significant difference ($p > 0.05$) compared with the control group (8.2 ± 1.14 times) (Figure 7a). When the concentration was higher than 8.4 μg/L, the number of molts began to decrease. The total number of molts in the group exposed to the highest concentration of 26.7 μg/L (6.6 ± 1.43 times) was significantly less than that of the control group ($p < 0.05$).

Except for the 8.4 μg/L exposure group, which had a slightly earlier first brood time than the control group (7.1 ± 0.316 days versus 7.2 ± 0.422 days), the first brood time in all other groups was delayed compared to the control group. In particular, the first brood time of *D. magna* in the 15 and 26.7 μg/L exposure groups (8.4 ± 2.297 and 8.75 ± 1.389 days, respectively) was significantly later than that of the control group ($p < 0.05$) (Figure 7b).

The number of first brood neonates and total offspring produced by *D. magna* showed an initial increase followed by a decrease with increasing MB concentration. The highest number of first brood neonates (19.3 ± 3.302) and total offspring produced (65.3 ± 9.21) were observed in the 8.4 μg/L concentration group, which was significantly higher than the control group ($p < 0.05$). In contrast, the lowest number of first brood neonates (8.875 ± 4.051) and total offspring produced (43.5 ± 4.75) was observed in the highest concentration group (26.7 μg/L), which was significantly lower than the control group and other exposure groups ($p < 0.05$; Figure 7c,d). The number of broods produced in all exposure groups was higher compared to the control group, and the number was significantly higher in the 8.4 and 26.7 μg/L groups than the control group ($p < 0.05$), with 4.1 ± 0.57 and 4.1 ± 0.35 broods, respectively (Figure 7e). The average number of neonates produced per brood decreased with increasing MB concentration, and all exposure groups had fewer neonates per brood than the control group. The lowest average number of neonates produced per brood was observed in the highest concentration group (26.7 μg/L), with 10.58 ± 1.23, which was significantly lower compared to the control group ($p < 0.05$; Figure 7f). The cumulative number of living offspring in each group over 21 days is shown in Figure 8.

In addition, "abortion" was observed during the experiment (Figure 9). One *D. magna* in the 15 μg/L exposure group aborted seven eggs on the seventh day. The undeveloped eggs fell off the brood chamber and adhered to the molted carapace. Another individual in the 26.7 μg/L exposure group aborted four eggs on the fifth day, and the eggs sank to the bottom of the beaker.

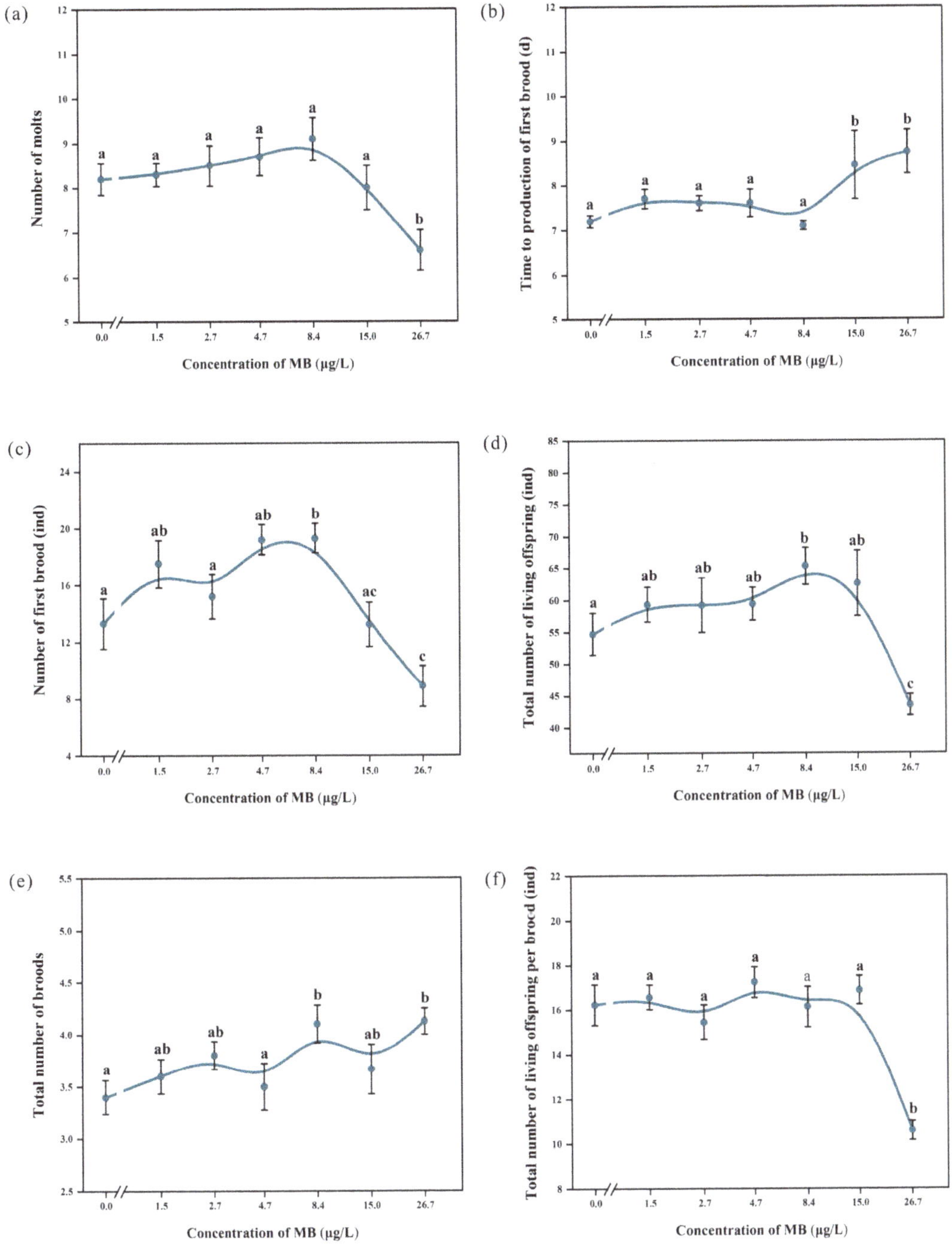

Figure 7. Influences of MB on (**a**) number of molts, (**b**) time to production of first brood, (**c**) number of first brood, (**d**) total number of living offspring, (**e**) total number of broods, and (**f**) total number of living offspring per brood (mean ± SE; $n = 10$). Significant differences ($p < 0.05$) among groups are indicated by different letters.

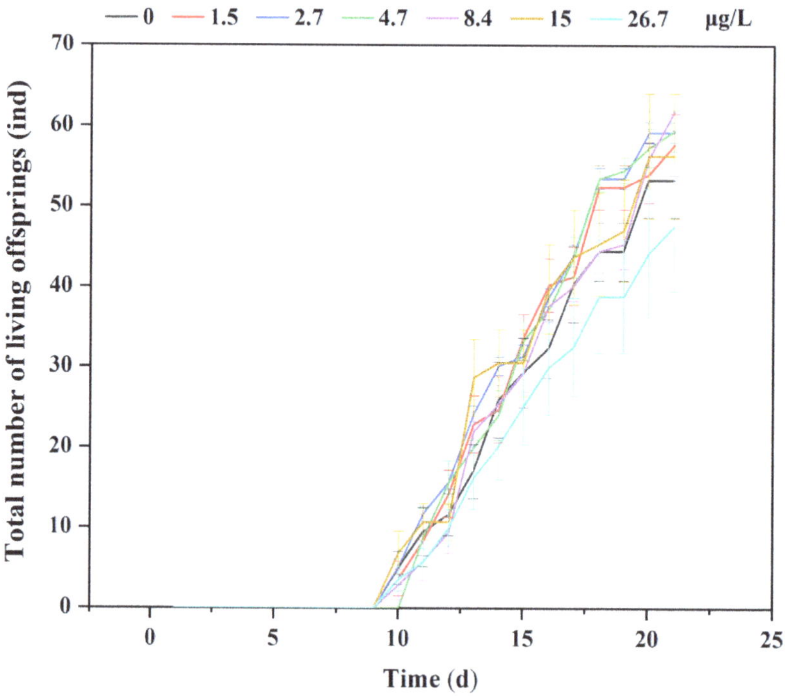

Figure 8. Cumulative number of living offspring over a 21-day period. Means and standard errors are shown.

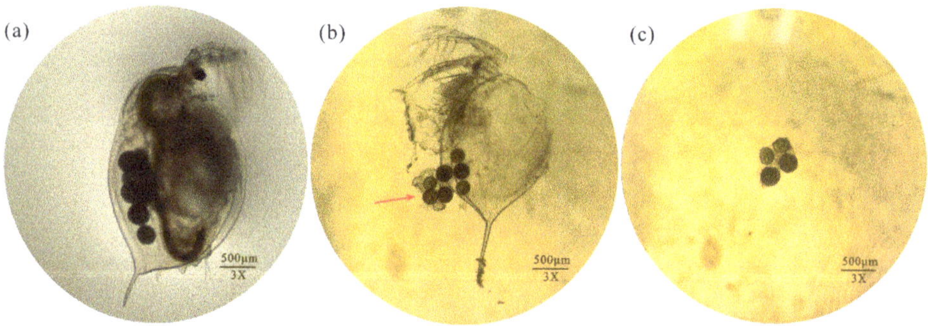

Figure 9. (**a**) Control group of *D. magna* and aborted eggs in groups exposed to (**b**) 15 and (**c**) 26.7 μg/L observed under a stereomicroscope (3×).

As the concentration of MB increased, the intrinsic growth rate showed a trend of first increasing and then decreasing (Figure 10). The intrinsic growth rate was higher in the groups exposed to 1.5, 2.7, 4.7, 8.4, and 15 μg/L of methylparaben than the control group, while the rate was lower in the 26.7 μg/L group than the control group, but the difference was not significant ($p > 0.05$).

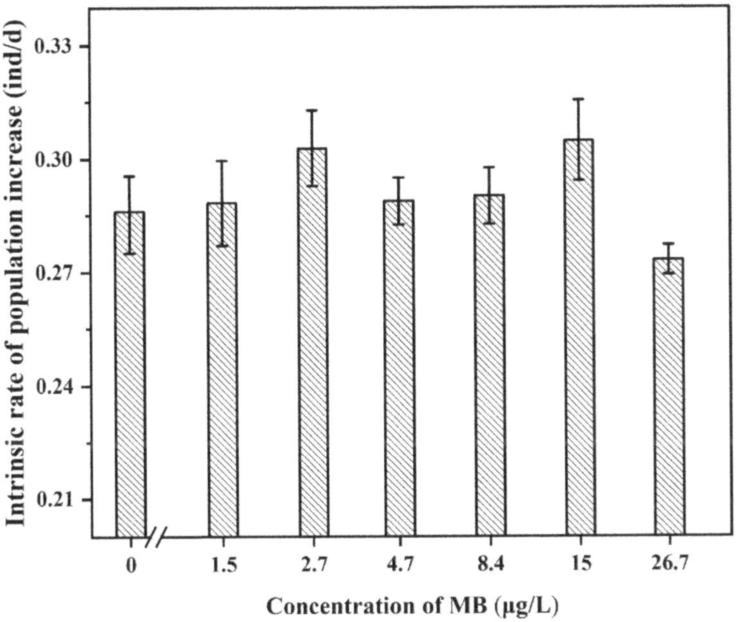

Figure 10. Intrinsic rate of population increase of *D. magna* (mean ± SE; $n = 10$).

3.5. Antioxidant Enzymes, Detoxification Enzymes, and Oxidative Damage Markers

The changes in superoxide dismutase (SOD) activity in *D. magna* tissues after exposure to MB are shown in Figure 11a. The SOD activity of the control group was the lowest, at 2.973 ± 0.479 U/mgprot. With increasing MB concentration, the SOD activity gradually increased. The SOD activity in *D. magna* exposed to 8.4 μg/L MB was significantly higher than that in the control group ($p < 0.05$). The SOD activity in *D. magna* exposed to 15 μg/L MB was 7.996 ± 0.878 U/mgprot, significantly higher than that in the control group ($p < 0.01$). When the MB concentration was 26.7 μg/L, the SOD activity of *D. magna* was the highest (15.497 ± 1.83 U/mgprot), which was 5.21 times that of the control group, and there was a significant difference ($p < 0.01$).

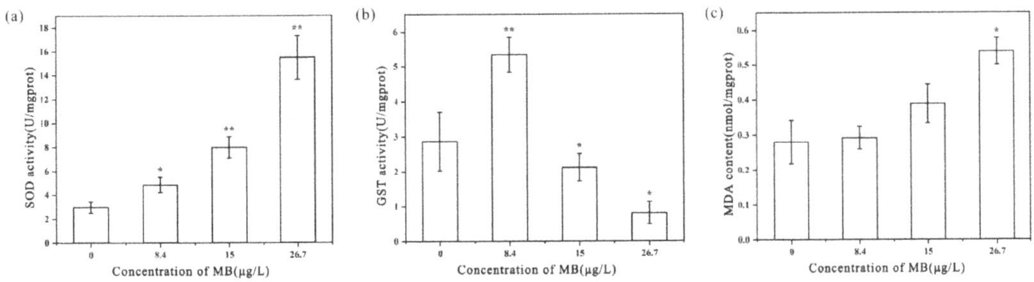

Figure 11. Effects of MB on (**a**) SOD activity, (**b**) GST activity, and (**c**) MDA content of *D. magna* (mean ± SE; $n = 3$; * $p < 0.05$ between the different treatment groups and the control group; ** $p < 0.01$ between the different treatment groups and the control group).

As the concentration of MB increased, GST activity in *D. magna* tissues was first induced and then inhibited. When the MB concentration was 8.4 μg/L, the GST activity was the highest (5.338 ± 0.505 U/mgprot), which was 1.86 times that in the control group (2.868 ± 0.835 U/mgprot), and the induced GST activity was significant ($p < 0.01$). However, when the MB concentration exceeded 8.4 μg/L, the GST activity significantly decreased compared to the control group ($p < 0.05$). This was evidenced by the GST activity in the *D. magna* tissues exposed to 15 and 26.7 μg/L MB, which was 2.119 ± 0.403 and 0.808 ± 0.319 c U/mgprot, respectively (Figure 11b).

As the concentration of MB increased, the MDA content in *D. magna* tissues also increased. However, there was no significant difference in MDA content between the 8.4 and 15 μg/L exposure groups and the blank control group ($p > 0.05$), with values of 0.292 ± 0.033 and 0.389 ± 0.055 nmol/mgprot, respectively. In contrast, the MDA content in *D. magna* exposed to the highest concentration of 26.7 μg/L (0.538 ± 0.038 nmol/mgprot) was 1.92 times higher than that in the control group (0.28 ± 0.062 nmol/mgprot) (Figure 11c).

4. Discussion

EC_{50} and LC_{50} are commonly used to assess the toxicity of pollutants to aquatic organisms. Previous studies reported 24 h-LC_{50} values of MB of 5.769 mg/L for *Litopenaeus vannamei* [32] and 31.60 mg/L for *Limnodrilus* [33], suggesting that the sensitivity to MB is highest in *D. magna* and that MB is a potent toxin for zooplankton. Furthermore, the EC_{50} and LC_{50} values vary depending on the type of pollutant. Abe et al. [30] reported EC_{50} values of two azo dyes, basic red 51 (BR51), a synthetic dye, and erythromycin (Ery), a natural dye, on *D. magna* of 0.10 mg/L (0.09–0.11) and 19.7 mg/L (15.7–24.9), respectively. Similarly, Verma [34] found that the 48 h-EC_{50} values of the azo dyes Remazol Parrot Green and Remazol Golden Yellow for *D. magna* were 55.32 and 46.84 mg/L, respectively, while Kanhere [35] reported that the 48 h-EC_{50} of malachite green was 0.77 mg/L. Compared with these results, *D. magna* was found to be more sensitive to the toxicity of methylene blue, a thiazide dye. Based on the United Nations Globally Harmonized System of Classification and Labelling of Chemicals (GHS Rev. 9, 2021) for acute aquatic toxicity (class I: 48 h-$EC_{50} \leq 1$ mg/L; class II: 1 mg/L < 48 h-$EC_{50} \leq 10$ mg/L; class III: 10 mg/L < 48 h-EC_{50}), the toxicity of MB to *D. magna* falls into class I.

The heart rate and beat frequency of the thoracic limbs of *D. magna* are important indicators of their overall health, reflecting the status of their feeding, respiration, metabolism, and endocrine system [36]. Several chemicals have been found to affect *D. magna*'s heart rate and beat frequency [37,38]. In this experiment, higher MB concentrations were found to decrease the heart rate and beat frequency of the thoracic limbs in *D. magna*. This may be due to the accumulation of MB with prolonged exposure, resulting in high toxicity and difficulty degrading it [39]. Long-term accumulation of MB in *D. magna* may lead to physiological abnormalities, decreased heart rate, and diminished eating capacity. Similar findings were reported by Eghan et al. [40], who observed time- and dose-dependent suppression of heart rate and thoracic limb beat frequency in *D. magna* exposed to acrylamide. Bownik et al. [41] also reported a time- and dose-dependent decrease in heart rate and thoracic limb beat frequency in response to ketoprofen, while procaine penicillin caused a concentration-dependent depression of heart rate and beat frequency [42]. *D. magna* has a myogenic heart, which means the myocardial contractions are not influenced by brain activity [43–46]. Pirtle et al. [45] suggested that the autonomic beating of *D. magna's* heart relies on hyperpolarization activating T-type calcium channels and cyclic nucleotide-gated ion channels. MB alters the activity of these ion channels, leading to changes in heart rate, which could explain the observed decrease in heart rate in *D. magna*. A decreased heart rate implies reduced O_2 and nutrient supply to cells [47], which impacts lymphatic blood circulation and immune response [48], thereby affecting the beat frequency of the thoracic limbs.

The thoracic limbs, which are the feeding organs of *D. magna*, are covered with bristles that filter food from the water and deliver it to the mouth. A decrease in the beat frequency of the thoracic limbs can affect the water filtration rate of *D. magna*, reducing their feeding capacity. Food intake is crucial for energy replenishment, and feeding capacity is closely related to individual growth and reproduction ability [49]. Moreover, decreased beat frequency of the thoracic limbs may be associated with the intestinal contents. On the one hand, it can be caused by an increase in particulate matter, including food [50,51]. Lari et al. [50] studied the effect of oil sands process-affected water (OPSW), a by-product of bitumen extraction, on *D. magna* and found that the exposed group exhibited a change in the color of the intestine from green to brown, a significantly higher density of algal cells compared to the control group, and a simultaneous decrease in the beat frequency of the thoracic limbs. On the other hand, exposure to dissolved toxicants can lead to decreased Na^+/K^+-ATPase activity in *D. magna*, resulting in reduced transmission between thoracic limb neurons and muscles and, in turn, decreased beat frequency of the thoracic limbs [36]. Additionally, blue residue was observed in the intestines of exposed *D. magna*, partially obstructing the flow of intestinal contents and potentially contributing to reduced feeding capacity.

Chronic exposure to MB significantly impacted the growth and reproductive capacity of *D. magna*. Specifically, the 26.7 μg/L exposure group exhibited significantly shorter body length and fewer molts than the control group, indicating that MB had some developmentally toxic effects, strongly inhibiting the growth and development of *D. magna*, as it normally develops through molting.

Regarding reproduction, exposure to high concentrations of MB caused a significant delay in the production of the first brood. At medium MB concentrations, the number of the first brood and total number of living offspring significantly increased, while at higher concentrations they decreased, indicating hormesis, a stimulative effect at low concentrations, and an inhibitory effect at high concentrations [52]. The total number of broods was higher in all exposure groups than in the control group, but the number of living offspring per brood was lower in the exposure groups. When considering the 21-day reproduction results, the increase in the number of the first brood and total number of living offspring in the 8.4 μg/L group was due to the earlier sexual maturation time, while the decrease in the 26.7 μg/L group was due to a significant delay in sexual maturation and a decrease in the number of living offspring per brood. This adaptive response to the environment aligns with previous studies, suggesting that *D. magna* deliberately increase the number of reproductions and reduce the number of single reproductions to stabilize the population in adverse environments [53].

Many studies have demonstrated the occurrence of hormesis in the growth, reproduction, and swimming behavior of *D. magna* when exposed to various contaminants. For instance, low concentrations of bioplastics promoted the reproductive rate, while higher concentrations inhibited it [54]. Similarly, low concentrations of ciprofloxacin and ofloxacin were found to stimulate a shortening of the first oogenesis time and an increase in brood size in *D. magna* [55]. In studies investigating the effects of fluoxetine and propranolol on *D. magna* swimming activity [56], intermediate drug doses (1~10 μg/L) significantly promoted swimming activity, while high doses (>100 μg/L) had the opposite effect, causing a significant decrease in swimming activity.

The decreased reproductive capacity of *D. magna* in the high-concentration groups may be attributed to the allocation of energy, primarily used for growth, reproduction, and basal metabolism. The stress response triggered by MB elevated the basal metabolic energy consumption of *D. magna*, thereby reducing the available energy reserves for growth and reproduction [57]. Consequently, there was a decline in reproductive ability and, in certain instances, even the occurrence of "abortion". These findings are consistent with those of a previous study [57].

The intrinsic rate of population increase is a measure of the population's ability to expand under ideal conditions. In this study, the intrinsic rate of population increase initially showed an increasing trend with increasing MB concentration, followed by a decreasing trend. However, these trends were not significantly different from those observed in the control group. The concentration of energy allocated to basal metabolism in *D. magna* under the stress of MB may influence its growth and reproduction, contributing to the observed pattern. The sensitivity of reproduction parameters to MB varied, and the order of sensitivity for these indicators was as follows: number of first brood = total number of broods = total number of living offspring > time to production of first brood > cumulative molts = number of living offspring per brood.

SOD plays a vital role in *D. magna*'s antioxidant defense system by eliminating reactive oxygen species (ROS) through the catalysis of superoxide anion radicals (O_2^-) into hydrogen peroxide (H_2O_2) and oxygen (O_2) [58]. This enzyme exerts a protective effect on the cells of the organism. Assessing SOD activity in *D. magna* can provide insights into its ability to adapt to MB exposure. In our study, SOD activity exhibited a continuous increase with escalating MB concentration. SOD activity in the highest concentration group was 5.21 times higher than that in the control group. These findings indicate that under the stress of MB, *D. magna* consistently enhanced its antioxidant capacity in response to the increasing assault of ROS and the unfavorable environment.

Excess reactive oxygen radicals in *D. magna* tissues that cannot be scavenged by SOD can lead to lipid peroxidation of cell membranes, causing cellular damage. This damage can be assessed by measuring the MDA content. In our study, the MDA content in *D. magna* exhibited a positive correlation with increasing MB concentration and was significantly higher compared to the control group, indicating oxidative damage. Despite the increased SOD activity, the MDA content also increased, suggesting that SOD was unable to fully eliminate all ROS. It is noteworthy that the highest MB concentration tested did not inhibit SOD activity, as reported in previous studies by Shen et al. [59] and Duan et al. [60] using different compounds but demonstrating similar effects. The elevated MDA levels in our study imply that MB exposure stimulated the generation of intracellular ROS in *D. magna*, leading to oxidative alterations of cellular components.

GST is an essential detoxifying enzyme that plays a crucial role in scavenging free radicals and facilitating detoxification processes. It catalyzes the conjugation of harmful endogenous or exogenous substances with reduced glutathione (GSH), forming more soluble and nontoxic derivatives that can be efficiently excreted or broken down by enzymes [61]. GST also possesses the ability to scavenge excess ROS, limit lipid peroxidation, and mitigate oxidative stress-induced damage [62]. In this study, GST activity exhibited hormetic effects, with induction observed in the low-concentration group and inhibition in the high-concentration group. At low concentrations of MB, GST activity increased to 1.86 times that of the control group, effectively scavenging free radicals and serving a detoxification function. However, at high MB concentrations, GST activity declined rapidly, possibly due to the depletion of intracellular GSH content. Consequently, toxins accumulated, disrupting the balance between free radical production and elimination and leading to the inactivation of GST, impairing its normal participation in the detoxification reaction. This finding is consistent with previous research [59], which demonstrated that after 24 h of dibutyl phthalate exposure, GST activity in *D. magna* neonates was dramatically elevated at 0.5 mg/L and reduced at 2 mg/L.

The influence of MB on *D. magna* involves complex physiological and biochemical processes, and our study examined only its effects on growth and reproduction. Consequently, there remains a knowledge gap concerning the toxicological mechanism of MB. Future investigations could bridge this gap by integrating conventional toxicological analysis with ecotoxicological genomics data, including transcriptomics, proteomics, metabolomics, and epigenomics. This comprehensive approach would enable a thorough exploration of the effects of MB on *D. magna* and other zooplankton species [63].

5. Conclusions

Methylene blue exhibited pronounced toxicity toward *D. magna*, with increasing toxic effects correlating with increasing concentration. Chronic exposure to MB significantly affected the growth and reproduction of *D. magna*, with heart rate and thoracic limb beat frequency proving to be more sensitive indicators than body length. Furthermore, MB could induce antioxidant stress in *D. magna*. The maximal concentration of MB at which no adverse effects were observed (NOEC) was determined to be 4.7 μg/L. Establishing water quality criteria for MB primarily relies on zooplankton, particularly *D. magna*, which is the most sensitive species to MB contamination (Supplementary Materials, Figure S2).

Supplementary Materials: The following supporting information can be downloaded at: https://www.mdpi.com/article/10.3390/toxics11070594/s1, Table S1. Acute toxicity of methylene blue to aquatic animals. Figure S1. Developmental stages of Daphnia magna under the stereomicroscope. Figure S2. Logistic fitting curve of species mean acute values (SMAV) of methylene blue. References [32,33,64–68] are cited in the supplementary materials.

Author Contributions: Conceptualization, G.J. and S.L.; data curation, G.J. and S.L.; formal analysis, S.L., G.J. and Y.C.; investigation, S.L., M.W. and Y.C.; methodology, S.L., M.W. and Y.C.; project administration, G.J.; resources, G.J.; visualization, G.J. and S.L.; writing—original draft, S.L.; writing—review and editing, M.W., Y.C. and G.J. All authors have read and agreed to the published version of the manuscript.

Funding: This work was supported by the program of the Shanghai Collaborative Innovation Center for Cultivating Elite Breeds and Green-culture of Aquaculture Animals (No. 2021-KJ-02-12), the Innovation Project for Chongming Agriculture Industry from Chongming District Agriculture Commission of Shanghai (2022CNKC-01-05), Shanghai Ocean University (A2-2006-21-200301), and Luo Zhaorao Science and Innovation Fund (A1-2004-22-201315).

Institutional Review Board Statement: Not applicable.

Informed Consent Statement: Not applicable.

Data Availability Statement: The data presented in this study are available upon request from the corresponding author.

Acknowledgments: The authors are grateful to Hongbo Pan, Wei Zhang, and Ruilei Zhang for their kind help with the experiment. They also thank the anonymous reviews.

Conflicts of Interest: The authors declare no conflict of interest.

References

1. Balarak, D.; Bazzi, M.; Shehu, Z.; Chandrika, K. Application of Surfactant-Modified Bentonite for Methylene Blue Adsorption from Aqueous Solution. *Orient. J. Chem.* **2020**, *36*, 293–299. [CrossRef]
2. Koyuncu, H.; Kul, A.R. Removal of methylene blue dye from aqueous solution by nonliving lichen (*Pseudevernia furfuracea* (L.) Zopf.), as a novel biosorbent. *Appl. Water Sci.* **2020**, *10*, 72. [CrossRef]
3. Mijinyawa, A.H.; Durga, G.; Mishra, A. A sustainable process for adsorptive removal of methylene blue onto a food grade mucilage: Kinetics, thermodynamics, and equilibrium evaluation. *Int. J. Phytoremed.* **2019**, *21*, 1122–1129. [CrossRef]
4. Safarik, I.; Safarikova, M. Detection of low concentrations of malachite green and crystal violet in water. *Water Res.* **2002**, *36*, 196–200. [CrossRef] [PubMed]
5. Van Tran, T.; Nguyen, D.T.; Kumar, P.S.; Din, A.T.; Qazaq, A.S.; Vo, D.V. Green synthesis of Mn_3O_4 nanoparticles using *Costus woodsonii* flowers extract for effective removal of malachite green dye. *Environ. Res.* **2022**, *214 Pt 2*, 113925. [CrossRef]
6. Sarojini, G.; Babu, S.V.; Rajamohan, N.; Rajasimman, M. Performance evaluation of polymer-marine biomass based bionanocomposite for the adsorptive removal of malachite green from synthetic wastewater. *Environ. Res.* **2022**, *204*, 112132. [CrossRef] [PubMed]
7. Majidian, S.; Taghavi, M.; Kohan, N.A.; Dehghan, A.; Afsharnia, M. Photocatalytic degradation of methylene blue dye using bismuth oxyiodide from aqueous solutions. *Int. J. Environ. Anal. Chem.* **2022**, 1–13. [CrossRef]
8. Meili, L.; Lins, P.; Costa, M.; Almeida, R.; Abud, A.; Soletti, J.; Dotto, G.; Tanabe, E.; Sellaoui, L.; Carvalho, S.; et al. Adsorption of methylene blue on agroindustrial wastes: Experimental investigation and phenomenological modelling. *Prog. Biophys. Mol. Biol.* **2019**, *141*, 60–71. [CrossRef]
9. Kosswattaarachchi, A.M.; Cook, T.R. Repurposing the Industrial Dye Methylene Blue as an Active Component for Redox Flow Batteries. *Chemelectrochem* **2018**, *5*, 3437–3442. [CrossRef]
10. Lawagon, C.P.; Amon, R.E.C. Magnetic rice husk ash 'cleanser' as efficient methylene blue adsorbent. *Environ. Eng. Res.* **2020**, *25*, 685–692. [CrossRef]
11. Zaini, M.A.A.; Sudi, R.M. Valorization of human hair as methylene blue dye adsorbents. *Green Process. Synth.* **2018**, *7*, 344–352. [CrossRef]

12. Zhou, S.; Du, Z.; Li, X.; Zhang, Y.; He, Y.; Zhang, Y. Degradation of methylene blue by natural manganese oxides: Kinetics and transformation products. *R. Soc. Open Sci.* **2019**, *6*, 190351. [CrossRef]
13. Perlberg, S.T.; Diamant, A.; Ofir, R.; Zilberg, D. Characterization of swim bladder non-inflation (SBN) in angelfish, Pterophyllum scalare (Schultz), and the effect of exposure to methylene blue. *J. Fish Dis.* **2008**, *31*, 215–228. [CrossRef]
14. Soltanian, S.; Gholamhosseini, A.; Banaee, M. Effects of exposure to a therapeutic level of methylene blue on antioxidant capacity, haemato-immunological responses and resistance of goldfish, *Carassius auratus* to *Aeromonas hydrophila*. *Aquac. Res.* **2021**, *52*, 2640–2650. [CrossRef]
15. Turnipseed, S.B.; Roybal, J.E.; Plakas, S.M.; Pfenning, A.P.; Hurlbut, J.A.; Long, A.R. Determination of methylene blue in channel catfish (*Ictalurus punctatus*) tissue by liquid chromatography with visible detection. *J. Aoac Int.* **1997**, *80*, 31–35. [CrossRef] [PubMed]
16. Pandey, S. A comprehensive review on recent developments in bentonite-based materials used as adsorbents for wastewater treatment. *J. Mol. Liq.* **2017**, *241*, 1091–1113. [CrossRef]
17. Pandey, S.; Do, J.Y.; Kim, J.; Kang, M. Fast and highly efficient removal of dye from aqueous solution using natural locust bean gum based hydrogels as adsorbent. *Int. J. Biol. Macromol.* **2019**, *143*, 60–75. [CrossRef]
18. Forro, L.; Korovchinsky, N.M.; Kotov, A.A. Global diversity of cladocerans (Cladocera; Crustacea) in freshwater. *Hydrobiologia* **2008**, *595*, 177–184. [CrossRef]
19. Dietrich, S.; Ploessl, F.; Bracher, F.; Laforsch, C. Single and combined toxicity of pharmaceuticals at environmentally relevant concentrations in *Daphnia magna*—A multigenerational study. *Chemosphere* **2010**, *79*, 60–66. [CrossRef]
20. Tatarazako, N.; Oda, S. The water flea *Daphnia magna* (Crustacea, Cladocera) as a test species for screening and evaluation of chemicals with endocrine disrupting effects on crustaceans. *Ecotoxicology* **2007**, *16*, 197–203. [CrossRef]
21. Fan, W.; Cui, M.; Liu, H.; Wang, C.; Shi, Z.; Tan, C.; Yang, X. Nano-TiO_2 enhances the toxicity of copper in natural water to *Daphnia magna*. *Environ. Pollut.* **2011**, *159*, 729–734. [CrossRef]
22. Reyes, V.P.; Ventura, M.A.; Amarillo, P.B. Ecotoxicological Assessment of Water and Sediment in Areas of Taal Lake with Heavy Aquaculture Practices Using Allium cepa and *Daphnia magna* Assay. *Philipp. J. Sci.* **2022**, *151*, 969–974. [CrossRef]
23. Day, K.; Kaushik, N.K. An assessment of the chronic toxicity of the synthetic pyrethroid, fenvalerate, to *Daphnia galeata* mendotae, using life tables. *Environ. Pollut.* **1987**, *44*, 13–26. [CrossRef] [PubMed]
24. Kim, H.Y.; Lee, M.J.; Yu, S.H.; Kim, S.D. The individual and population effects of tetracycline on *Daphnia magna* in multigenerational exposure. *Ecotoxicology* **2012**, *21*, 993–1002. [CrossRef]
25. Sanpradit, P.; Peerakietkhajorn, S. Disturbances in growth, oxidative stress, energy reserves and the expressions of related genes in *Daphnia magna* after exposure to ZnO under thermal stress. *Sci. Total Environ.* **2023**, *21*, 161682. [CrossRef]
26. Mahaye, N.; Musee, N. Effects of Two Antiretroviral Drugs on the Crustacean *Daphnia magna* in River Water. *Toxics* **2022**, *10*, 423. [CrossRef] [PubMed]
27. Galhano, V.; Zeumer, R.; Monteiro, M.S.; Knopf, B.; Meisterjahn, B.; Soares, A.M.; Loureiro, S.; Schlechtriem, C.; Lopes, I. Effects of wastewater-spiked nanoparticles of silver and titanium dioxide on survival, growth, reproduction and biochemical markers of *Daphnia magna*. *Sci. Total Environ.* **2022**, *839*, 156079. [CrossRef]
28. Yee, M.S.; Khiew, P.S.; Chiu, W.S.; Tan, Y.F.; Leong, C.O. Polyethyleneimine-Capped Silver Nanoparticles as Antifouling Photocatalyst for Wastewater Treatment. *Mater. Today Proc.* **2019**, *19*, 1497–1506.
29. OECD. *Test No. 211: Daphnia Magna Reproduction Test*; OECD: Paris, France, 2012.
30. Abe, F.R.; Machado, A.L.; Soares, A.M.; de Oliveira, D.P.; Pestana, J.L. Life history and behavior effects of synthetic and natural dyes on Daphnia magna. *Chemosphere* **2019**, *236*, 124390. [CrossRef]
31. Hu, H.Y.; Xi, Y.L. Demographic parameters and mixis of three *Brachionus angularis Gosse* (Rotatoria) strains fed on different algae. *Limnologica* **2008**, *38*, 56–62. [CrossRef]
32. Zhou, G.; Rongrong, M.; Zongying, Y.; Kun, H. Acute toxicity of prometryn, phoxim, and methylene blue and their histopathological effects on *Penaeus vannamei*. *Asian J. Ecotoxicol.* **2020**, *15*, 279–289.
33. Yin, H.; Kan, X.; Shi, X.; Fan, X. The acute poisoning function of methyl to *Limnodilus hoffineisteri*. *Nat. Sci. J. Harbin Norm. Univ.* **2007**, *4*, 80–82.
34. Verma, Y. Acute toxicity assessment of textile dyes and textile and dye industrial effluents using *Daphnia magna* bioassay. *Toxicol. Ind. Health* **2008**, *24*, 491–500. [CrossRef] [PubMed]
35. Kanhere, J.; Gopinathan, R.; Banerjee, J. Cytotoxicity and Genotoxicity of Malachite Green on Non-Target Aquatic Organisms: *Chlorella Pyrenoidosa* and *Daphnia magna*. *Water Air Soil Pollut.* **2014**, *225*, 1–8. [CrossRef]
36. Farkas, A.; Somogyvári, D.; Kovács, A.W.; Mörtl, M.; Székács, A.; Győri, J. Physiological and metabolic alterations induced by commercial neonicotinoid formulations in *Daphnia magna*. *Ecotoxicology* **2022**, *31*, 415–424. [CrossRef] [PubMed]
37. Steinkey, D.; Lari, E.; Woodman, S.G.; Steinkey, R.; Luong, K.H.; Wong, C.S.; Pyle, G.G. The effects of diltiazem on growth, reproduction, energy reserves, and calcium-dependent physiology in *Daphnia magna*. *Chemosphere* **2019**, *232*, 424–429. [CrossRef]
38. Wei, X.; Li, X.; Liu, H.; Lei, H.; Sun, W.; Li, D.; Dong, W.; Chen, H.; Xie, L. Altered life history traits and transcripts of molting- and reproduction-related genes by cadmium in Daphnia magna. *Ecotoxicology* **2022**, *31*, 735–745. [CrossRef]
39. Alencar, J.M.; Oliveira, F.J.; Airoldi, C.; Filho, E.C.S. Organophilic nickel phyllosilicate for reactive blue dye removal. *Chem. Eng. J.* **2014**, *236*, 332–340. [CrossRef]

40. Eghan, K.; Lee, S.; Kim, W.-K. Cardiotoxicity and neurobehavioral effects induced by acrylamide in Daphnia magna. *Ecotoxicol. Environ. Saf.* **2022**, *242*, 113923. [CrossRef]
41. Bownik, A.; Jasieczek, M.; Kosztowny, E. Ketoprofen affects swimming behavior and impairs physiological endpoints of *Daphnia magna*. *Sci. Total Environ.* **2020**, *725*, 138312. [CrossRef]
42. Bownik, A.; Ślaska, B.; Bochra, J.; Gumieniak, K.; Gałek, K. Procaine penicillin alters swimming behaviour and physiological parameters of Daphnia magna. *Environ. Sci. Pollut. Res.* **2019**, *26*, 18662–18673. [CrossRef]
43. Bekker, J.M.; Krijgsman, B.J. Physiological investigations into the heart function of *Daphnia*. *J. Physiol.* **1951**, *115*, 249–257. [CrossRef]
44. Campbell, A.K.; Wann, K.T.; Matthews, S.B. Lactose causes heart arrhythmia in the water flea Daphnia pulex. *Comp. Biochem. Physiol. Part B Biochem. Mol. Biol.* **2004**, *139*, 225–234. [CrossRef]
45. Pirtle, T.J.; Carr, T.L.; Khurana, T.; Meeker, G. ZD7288 and mibefradil inhibit the myogenic heartbeat in Daphnia magna indicating its dependency on HCN and T-type calcium ion channels. *Comp. Biochem. Physiol. Part A Mol. Integr. Physiol.* **2018**, *222*, 36–42. [CrossRef] [PubMed]
46. Yamagishi, H.; Ando, H.; Makioka, T. Myogenic Heartbeat in the Primitive Crustacean *Triops longicaudatus*. *Biol. Bull.* **1997**, *193*, 350–358. [CrossRef] [PubMed]
47. Jeong, T.Y.; Yoon, D.; Kim, S.; Kim, H.Y.; Kim, S.D. Mode of action characterization for adverse effect of propranolol in *Daphnia magna* based on behavior and physiology monitoring and metabolite profiling. *Environ. Pollut.* **2018**, *233*, 99–108. [CrossRef] [PubMed]
48. Bownik, A.; Kowalczyk, M.; Bańczerowski, J. Lambda-cyhalothrin affects swimming activity and physiological responses of Daphnia magna. *Chemosphere* **2018**, *216*, 805–811. [CrossRef]
49. Lu, G.; Yang, H.; Xia, J.; Zong, Y.; Liu, J. Toxicity of Cu and Cr Nanoparticles to Daphnia magna. *Water Air Soil Pollut.* **2016**, *228*, 18. [CrossRef]
50. Lari, E.; Steinkey, D.; Morandi, G.; Rasmussen, J.B.; Giesy, J.P.; Pyle, G.G. Oil sands process-affected water impairs feeding by Daphnia magna. *Chemosphere* **2017**, *175*, 465–472. [CrossRef]
51. Lovern, S.B.; Strickler, J.R.; Klaper, R.D. Behavioral and physiological changes in *Daphnia magna* when exposed to nanoparticle suspensions (titanium dioxide, nano-C_{60}, and $C_{60}H_xC_{70}H_x$). *Environ. Sci. Technol.* **2007**, *12*, 4465–4470. [CrossRef]
52. Calabrese, E.J.; Baldwin, L.A. Toxicology rethinks its central belief. *Nature* **2003**, *421*, 691–692. [CrossRef] [PubMed]
53. Lyu, K.; Cao, H.; Chen, R.; Wang, Q.; Yang, Z. Combined effects of hypoxia and ammonia to Daphnia similis estimated with life-history traits. *Environ. Sci. Pollut. Res.* **2013**, *20*, 5379–5387. [CrossRef] [PubMed]
54. Boisseaux, P.; Hopkinson, P.; Santillo, D.; Smith, C.; Garmulewicz, A.; Powell, Z.; Galloway, T. Environmental safety of second and third generation bioplastics in the context of the circular economy. *Ecotoxicol. Environ. Saf.* **2023**, *256*, 114835. [CrossRef] [PubMed]
55. Nguyen, T.-D.; Itayama, T.; Ramaraj, R.; Iwami, N.; Shimizu, K.; Dao, T.-S.; Pham, T.-L.; Maseda, H. Chronic ecotoxicology and statistical investigation of ciprofloxacin and ofloxacin to Daphnia magna under extendedly long-term exposure. *Environ. Pollut.* **2021**, *291*, 118095. [CrossRef]
56. Nielsen, M.E.; Roslev, P. Behavioral responses and starvation survival of Daphnia magna exposed to fluoxetine and propranolol. *Chemosphere* **2018**, *211*, 978–985. [CrossRef]
57. Sancho, E.; Villarroel, M.; Andreu, E.; Ferrando, M. Disturbances in energy metabolism of Daphnia magna after exposure to tebuconazole. *Chemosphere* **2009**, *74*, 1171–1178. [CrossRef]
58. Jemec, A.; Tišler, T.; Erjavec, B.; Pintar, A. Antioxidant responses and whole-organism changes in *Daphnia magna* acutely and chronically exposed to endocrine disruptor bisphenol A. *Ecotoxicol. Environ. Saf.* **2012**, *86*, 213–218. [CrossRef]
59. Shen, C.; Wei, J.; Wang, T.; Wang, Y. Acute toxicity and responses of antioxidant systems to dibutyl phthalate in neonate and adult *Daphnia magna*. *PeerJ* **2019**, *7*, e6584. [CrossRef]
60. Duan, S.; Fu, Y.; Dong, S.; Ma, Y.; Meng, H.; Guo, R.; Chen, J.; Liu, Y.; Li, Y. Psychoactive drugs citalopram and mirtazapine caused oxidative stress and damage of feeding behavior in *Daphnia magna*. *Ecotoxicol. Environ. Saf.* **2022**, *230*, 113147. [CrossRef]
61. Coleman, J.; Blake-Kalff, M.; Davies, E. Detoxification of xenobiotics by plants: Chemical modification and vacuolar compartmentation. *Trends Plant Sci.* **1997**, *2*, 144–151. [CrossRef]
62. Li, W.; Zhu, L.; Du, Z.; Li, B.; Wang, J.; Wang, J.; Zhang, C.; Zhu, L. Acute toxicity, oxidative stress and DNA damage of three task-specific ionic liquids ([C_2NH_2MIm] BF_4,[MOEMIm] BF_4, and [HOEMIm] BF_4) to zebrafish (*Danio rerio*). *Chemosphere* **2020**, *249*, 126119. [CrossRef] [PubMed]
63. Kim, H.J.; Koedrith, P.; Seo, Y.R. Ecotoxicogenomic Approaches for Understanding Molecular Mechanisms of Environmental Chemical Toxicity Using Aquatic Invertebrate, *Daphnia* Model Organism. *Int. J. Mol. Sci.* **2015**, *16*, 12261–12287. [CrossRef] [PubMed]
64. Dai, Y.; Xie, N.; Ma, H.; Dai, Y.; Xu, B.; Lin, Q.; Huang, H. Acute toxicity test of copper iron mixture, metrifonate, methylene blue and povidone-iodine on summerlings of *Megalobrama terminalis*. *Fish. Sci. Technol. Inf.* **2020**, *47*, 4.
65. Chen, W.; Qu, J.; Zhou, F. Study on the acute toxicity of methylene blue to young *Rhodeus Ocellatus*. *Guangdong Agric. Sci.* **2010**, *37*, 131–132.
66. Rifici, L.M.; Cherry, D.S.; Farris, J.L.; Cairns, J., Jr. Acute and subchronic toxicity of methylene blue to larval fathead minnows (*Pimephales promelas*): Implications for aquatic toxicity testing. *Environ. Toxicol. Chem.* **1996**, *15*, 1304–1308. [CrossRef]

67. Hanks, K.S. Toxicity of some chemical therapeutics to the commercial shrimp, *Penaeus californiensis*. *Aquaculture* **1976**, *7*, 293–294. [CrossRef]
68. Willford, W.A.; USDOI, FWS, Bur Sport Fish Wildl. Invest Fish Control No.18, Resourc Publ No. 35: 10 (1966) as Cited in the ECOTOX Database. 28 November 2022.

toxics

MDPI

Article

Toxicity of Difenoconazole and Atrazine and Their Photodegradation Products on Aquatic Biota: Environmental Implications in Countries Lacking Good Agricultural Practices

Julia Mendieta Herrera [1], Carlos Iñiguez Armijos [2,*], Daniel Rosado Alcarria [3] and Silvio Aguilar Ramírez [1]

[1] Departamento de Química y Ciencias Exactas, Universidad Técnica Particular de Loja, San Cayetano Alto s/n, Loja 1101608, Ecuador

[2] Departamento de Ciencias Biológicas y Agropecuarias, Universidad Técnica Particular de Loja, San Cayetano Alto s/n, Loja 1101608, Ecuador

[3] Department of Hydrology and Water Resources Management, Institute for Natural Resource Conservation, Kiel University, 24118 Kiel, Germany

* Correspondence: cainiguez@utpl.edu.ec

Abstract: Agriculture is fundamental for human development, but it may also have a range of unwanted effects on ecosystems when pesticides inadvertently enter the environment. We determined the toxicity of difenoconazole and atrazine, as well as their photodegradation products, on the bioindicators *Lemna minor* and *Daphnia magna*. For *L. minor*, we assessed the number of leaves, biomass, and chlorophyll content exposed to different concentrations of difenoconazole (0–8 mg/L) and atrazine (0–3.84 mg/L). For *D. magna*, we assessed the mortality to difenoconazole (0–1.6 mg/L) and atrazine (0–80 mg/L). We found that the higher the concentrations of the pesticides, the higher the toxicity for both bioindicators. In *L. minor*, the highest toxicity for atrazine was 0.96 mg/L, whereas for difenoconazole, it was 8 mg/L. For *D. magna*, the 48 h LC_{50} for difenoconazole was 0.97 mg/L, while for atrazine, it was 86.19 mg/L. For *L. minor*, the toxicity of difenoconazole and atrazine was not different compared to that of their photodegradation products. In contrast, for *D. magna*, difenoconazole, but not atrazine, was more toxic compared to its respective photodegradation products. Pesticides are a serious threat to aquatic biota, and their photodegradation products remain toxic in the environment. Additionally, the use of bioindicators can help monitor these pollutants in aquatic ecosystems in countries where the application of pesticides is imperative for agricultural production.

Keywords: *Lemna minor*; *Daphnia magna*; pesticides; aquatic toxicity

1. Introduction

The contamination of aquatic ecosystems worldwide is a drastic issue, which worsens every day by the inadvertent entry of pesticides into the environment [1]. Only very low percentages of applied pesticides reach the target plant, and most pesticides simply end up in aquatic ecosystems through percolation, evaporation, leaching, runoff, and erosion [2]. Thus, it is almost impossible to trace the flow of pesticides in the environment.

Difenoconazole and atrazine stand out as being among the most used pesticides in agriculture that have been detected as contaminants in the environment [3,4]. Difenoconazole (triazole family) is a fungicide that interferes with the biosynthesis of ergosterol in fungi, acting mainly on the demethylation of C^{14}, which causes morphological and functional alterations of the cell wall [4]. In addition, it belongs to the group of endocrine disruptors, which are known to cause damage to human health and cause acute and chronic toxicity in aquatic environments (H400 and H410, respectively, according to the classification of the Globally Harmonized System of Classification and Labelling of Chemical Products (GHS)) [5,6]. Atrazine (triazine family) is a herbicide that inhibits photosynthetic electron transport in leaves [4]. It is leached through the soil by rain or irrigation water until it reaches bodies of water, where it is frequently detected due to its low solubility in water [5]. In addition, atrazine is listed as hazardous [6] (acute and chronic toxicity, H400 and 410, respectively) for aquatic environments.

Pesticides present in the environment are degraded by physicochemical (e.g., photodegradation by solar irradiation) or biological processes (e.g., degradation by microbial activity), giving rise to transformation products. In many cases, these transformation products have unknown effects on the environment [7], are known to cause loss of species [7], or have been classified as carcinogenic, neurotoxic, and teratogenic [8]. Some authors, such as Man et al. [9], found that photodegradation products of difenoconazole are toxic for fish but not for crustaceans. Klementová et al. [3] reported no toxicity of the combined photodegraded products of atrazine on aquatic plants, while crustaceans only suffered toxic effects after long-term exposure to atrazine. Evgenidou and Fytianos [10] investigated the transformation products of atrazine via different degradation forms. However, it is still necessary to understand how these compounds affect water bodies, their degradation pathways, the means of detection, the effects of chronic exposure, and the responses of aquatic biota.

Toxicity bioassays with biological indicators are already standardized techniques (Standard Methods 8211), which tests give reliable and reproducible/repeatable results [11]. The use of biological indicators in toxicity tests is on the rise due to the ease of installation, maintenance, and adaptation to laboratory conditions, which lowers the costs of studies [12]. *Lemna minor* is a floating macrophytic aquatic plant that inhabits freshwater bodies; it is commonly used in aquatic ecotoxicity tests and is recommended for toxicity evaluations in processes where pesticides are used [13,14]. Within the animal kingdom, *Daphnia magna* is the most commonly used cladoceran crustacean in ecotoxicological tests because it is easy to establish in the laboratory and has a short life cycle [15]. The USEPA [11,16] recommend the use of these species for the evaluation of agrochemical toxicity in aquatic environments.

Agricultural production in Ecuador contributes 8% to the country's total annual production [17]. According to the FAOSTAT data, Ecuador registered pesticide use of approximately 14.03 kg/ha in 2019 [18]. Difenoconazole is applied to control Black Sigatoka caused by the fungus *Mycospharella fijensis* in banana crops, and it is an agrochemical considered to be moderately dangerous (II) according to the toxicological category of the World Health Organization [8,17,19]. Atrazine, on the other hand, is applied as a weed controller in corn and sugar cane crops and it is an agrochemical considered to be slightly dangerous (III) [10,20]. In Ecuador, the areas planted with bananas occupy 165,080 ha, and the areas planted with sugarcane occupy more than 157,900 ha [17].

Therefore, the present study investigated the toxic effects of difenoconazole and atrazine and the potential toxicity of their photodegradation products on aquatic biota by using *L. minor* and *D. magna* as bioindicators. The photodegradation products were obtained by exposing both pesticides to UV irradiation in the laboratory. As the response variables, we used the number of leaves, biomass, and chlorophyll content of *L. minor* and percentage of mortality of *D. magna*. We expected that the toxicity of the photodegradation products is lower than that of the pesticides.

2. Materials and Methods

2.1. Difenoconazole and Atrazine Photodegradation

We used commercial-grade difenoconazole (Score® 250 EC, Syngenta S.A., Cartagena, Colombia) and commercial-grade atrazine (ATRAPAC® 900, Agripac, Guayaquil, Ecuador). The photodegradation products were obtained by exposing both pesticides to irradiation with a 6 W UV-C light lamp (200 to 280 nm) in a cylindrical reactor (Microfilter Ultraviolet Sterilization Filter model OPP-625 1.0 GPM. Sejong, Republic of Korea) with an adjustable flow rate peristaltic pump (Cole Parmer model 7523-80 Masterflex L/S, Waltham, MA, USA).

The photodegradation conditions for both pesticides were as follows: for difenoconazole, a 8 mg/L solution was prepared and then irradiated for 2 and 4 min, whereas for atrazine, a 3.84 mg/L solution was prepared and then irradiated for 15 min until photodegradation products were obtained (Table 1). Pesticide concentrations and exposure times (irradiation with UV-C light lamp) were established after performing and monitoring previous tests until photodegradation products were detected through high performance liquid chromatography (HPLC).

All tests during the assays were carried out with the previously mentioned commercial pesticides; for their analytical determination, we used the standard solutions for difenoconazole (Difenoconazol PESTANAL®, Sigma Aldrich, St. Louis, MO, USA) and atrazine (Atrazine PESTANAL®, Sigma Aldrich, St. Louis, MO, USA).

The analytical determination of difenoconazole, atrazine, and their photodegradation products was performed using a HPCL coupled to a mass spectrometer (Bruker amaZon Ion Trap Mass, Bremen, Germany) and a UV detector (Thermo Scientific, Dionex UltiMate 3000 modular system, Waltham, MA, USA).

2.2. Toxicity Tests with Lemna Minor

The toxicity tests for *L. minor* were carried out according to the guidelines of the Standard Methods 8211 [11]. Prior to the tests, individuals of *L. minor* (adapted and maintained under laboratory conditions [11]) were cultured for two weeks in a nutrient solution for duckweed (see Table S1 for details about solution) and simultaneously subjected to an adaptation period with the light (24 h; 2150–4300 lux) and temperature (24 ± 2 °C) conditions used during the toxicity tests. We prepared six concentrations of difenoconazole (0, 0.5, 1, 2, 4, and 8 mg/L) from a stock solution of difenoconazole (1000 mg/L), and six concentrations of atrazine (0, 0.12, 0.24, 0.48, 0.96, 1.92, and 3.84 mg/L) from a stock solution of atrazine (192 mg/L). To test the effect of the photodegradation products from both pesticides, individuals of *L. minor* were exposed to a solution of 8 mg/L of difenoconazole and to a solution of 3.84 mg/L irradiated with UV-C light. The selection of these concentrations was based on previous range-finder tests, where both bioindicators were exposed to different concentrations of each pesticide until we detected damage or alteration in the exposed organisms to select a reference initial concentration. Then, from the initial concentration, we successively increased the concentrations by two-fold.

Three replicates were created for each concentration of the pesticides and their photodegradation products. In each replicate, we placed 10 healthy individuals of *L. minor* in a 350 mL glass bowl containing 100 mL of the nutrient solution and an aliquot of each concentration. *L. minor* individuals were similar in size and composed of 2 to 3 leaves; they were exposed to the pesticides and photodegradation products for 7 days. During the exposure period, the medium was not refreshed. Once the exposure time elapsed, the individuals of *L. minor* were carefully removed and washed with DI water to determine the average total number of leaves, the average total biomass, and the average total chlorophyll content per replicate, as explained below.

The total number of leaves was counted at the beginning and at the end of the exposure time (7 days). A new leaf was determined as every shoot that was observed in the individuals. Any sign of chlorosis or deterioration in the leaves was also recorded. Then, the number of leaves and the total biomass were determined. The individuals were placed in 1.5 mL Eppendorf tubes that had been previously waxed and covered with parafilm, making 3 to 4 perforations. Then, the individuals were lyophilized for a period of 3 hours at −50 °C and 0.250 mBar (Labconco modelo 7754047. Kansas City, MO, USA) and weighed to the nearest 0.0001 mg. Once weighed, the individuals of *L. minor* were placed in centrifugation tubes (10 mL) with 2 mL of 80% acetone. Subsequently, they were placed in a water bath at 25 °C in the dark for 24 h, with continuous agitation for the extraction of chlorophyll Then, the chlorophyll concentration was quantified in a spectrophotometer (HACH DR 2800, Düsseldorf, Germany) by measuring the absorbance at 665 nm (Chl-*a*) and at 649 nm (Chl-*b*), respectively. We chose these absorbance values based on the literature [21,22] and on previous tests carried out in the laboratory after the maximum absorbance was determined.

2.3. Toxicity Tests with Daphnia Magna

The toxicity tests for *D. magna* were carried out based on the protocols of the Standard Methods 8711. Prior to the tests, individuals of *D. magna* were cultured for four weeks in a medium consisting of reconstituted hard water according to the Standard Methods 8010:1 (Table S2). During this time, the individuals of *D. magna* (adapted and maintained in the laboratory facilities [11]) were adapted to the light (16 h; 528–1076 lux) and temperature (20 ± 2 °C) conditions used in the toxicity tests. We prepared six concentrations of difenoconazole (0, 0.1, 0.2, 0.4, 0.8, and 1.6 mg/L) from a stock solution of difenoconazole (10 mg/L). For atrazine, we also prepared six concentrations (0, 10, 20, 40, 60, and 80 mg/L) from a stock solution of atrazine (500 mg/L). To test the effect of the photodegradation products from both pesticides, the individuals of *D. magna* were exposed to a solution of 1.6 mg/L of difenoconazole and to a solution of 80 mg/L of atrazine irradiated with UV-C light. The selection of these concentration was similarly made as explained above.

For each concentration of the pesticides and their photodegradation products, three replicates were created by placing 10 neonates (i.e., 24 h after hatching) of *D. magna* in a 250 mL beaker containing 50 mL of reconstituted hard water plus the aliquot of each concentration per replicate. The *D. magna* neonates were exposed to the pesticides and photodegradation products since hatching for 48 h. Once the exposure time was completed, the living and immobile individuals were quantified through observation. Then, immobility was used as a proxy of mortality, and its percentage was calculated.

2.4. Data Analysis

Data analysis was performed in the R programming environment version 4.1.3 [23]. A GLM was applied to evaluate the effects of the pesticides on *L minor* and *D. magna*. The total number of leaves, the total biomass, and the total chlorophyll of *L. minor*, and the percentage of mortality of *D. magna*, were used as the response variables against each concentration (treatment) of the pesticides. In another GLM, the effects of the pesticides and their photodegradation products were evaluated using the same response variables against the type of pesticide (pesticide vs. by-products) and the different concentrations (control vs. the highest concentration of each pesticide and by-product). The GLMs were fitted assuming an adequate error distribution using the *stats* package [23], i.e., Poisson error distribution for number of leaves, Gaussian error distribution for biomass and total chlorophyll, and binomial error distribution for percentage of mortality. Significant differences were evaluated by a pairwise comparison test using the *emmeans* package [24].

For the mortality percentage of *D. magna* only, the lethal concentrations (LC_{50}, LC_{20}, and LC_{10}) for difenoconazole and atrazine were calculated using the *drc* package [25]. The lethal concentrations were calculated with a 95% confidence interval.

Table 1. Summary of the characteristics and effects of the studied pesticides as well as their photodegradation products after UV irradiation. *m/z* refers to the observed mass/charge number ratio. RT indicates the retention time expressed in minutes.

Name	Structure	Formula	*m/z*	RT	Family	Effect
Difenoconazole	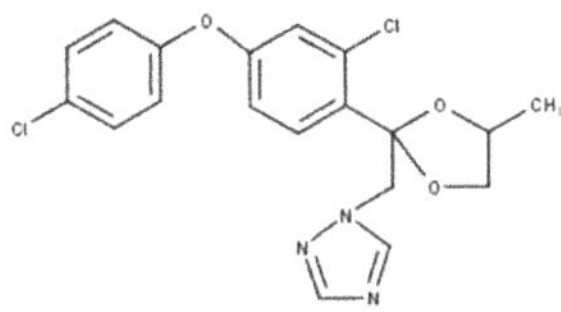	$C_{19}H_{17}Cl_2N_3O_3$	405.9	6.9	Triazole	Inhibitor of the biosynthesis of ergosterol in the cellular membrane of fungi [4].
A-Difenoconazole	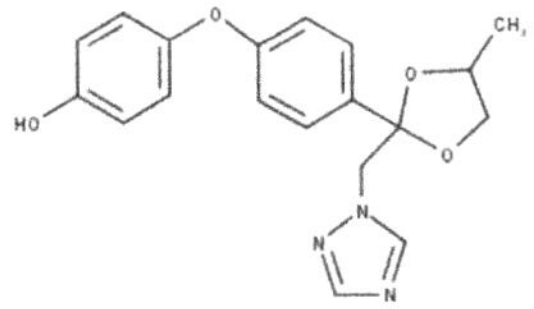	$C_{19}H_{19}N_3O_4$	354.9	2.9	Triazole	Affects the normal development of aquatic vegetation and reduces chlorophyll production. Affects the mortality of invertebrates (in this study).
B-Difenoconazole	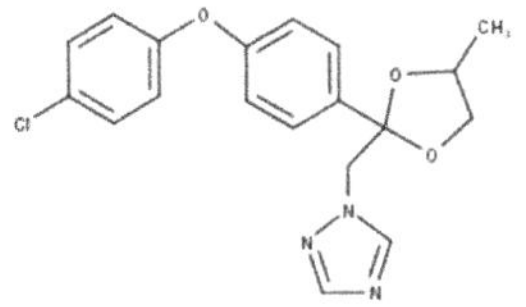	$C_{19}H_{18}ClN_3O_3$	370.0	5.2	Triazole	Affects the normal development of aquatic vegetation and reduces chlorophyll production. Affects the mortality of invertebrates (in this study).
Atrazine	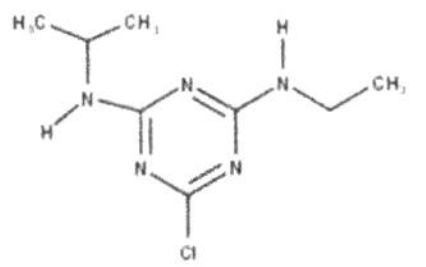	$C_8H_{14}ClN_5$	215.7	19.0	Triazine	Inhibitors of photosynthetic electron transport [26].
A-Atrazine	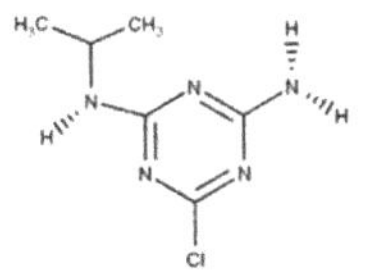	$C_6H_{10}ClN_5$	187.0	5.5	Triazine	Affects the normal development of aquatic vegetation and reduces chlorophyll production (in this study).

3. Results

3.1. Photodegradation of Difenoconazole and Atrazine

After the photodegradation of the pesticides, we identified two photodegradation products for difenoconazole (A-Difenoconazole and B-Difenoconazole) and one for atrazine (A-Atrazine), as shown in Table 1. For A-difenoconazole and B-Difenoconazole, the molecular formulas are $C_{19}H_{19}N_3O_4$ and $C_{19}H_{18}ClN_3O_3$, respectively, while for A-Atrazine, the molecular formula is $C_6H_{10}ClN_5$. The three photodegradation products were identified at the following retention times: A-Difenoconazole at 3 min, B-Difenoconazole at 5 min, and A-Atrazine at 5.5 min (Table 1).

3.2. Toxicity Tests with *L. minor*

Both difenoconazole and atrazine have a significant effect on *L. minor* (Table 2). The number of leaves decreases significantly as the concentrations of both pesticides increase, as shown in Figure 1. For difenoconazole, the highest number of leaves is observed in the control, which is different from the other concentrations, while there are no significant differences in the number of leaves between the concentrations of 0.5 and 1 mg/L and between the concentrations of 2, 4, and 8 mg/L (Figure 1). Regarding biomass, the highest biomass is observed in the control, and there are no significant differences between the concentrations of 0.5 and 1 mg/L; however, there are significant differences for the higher concentrations, with the concentration of 8 mg/L of difenoconazole having a greater effect on the biomass of *L. minor* (Figure 1). The total content of chlorophyll is significantly higher in the control and in the concentration of 0.5 mg/L of difenoconazole, and it decreases with higher concentrations, with the concentration of 8 mg/L of difenoconazole being the dose with the greatest effect on the chlorophyll content of *L. minor* (Figure 1).

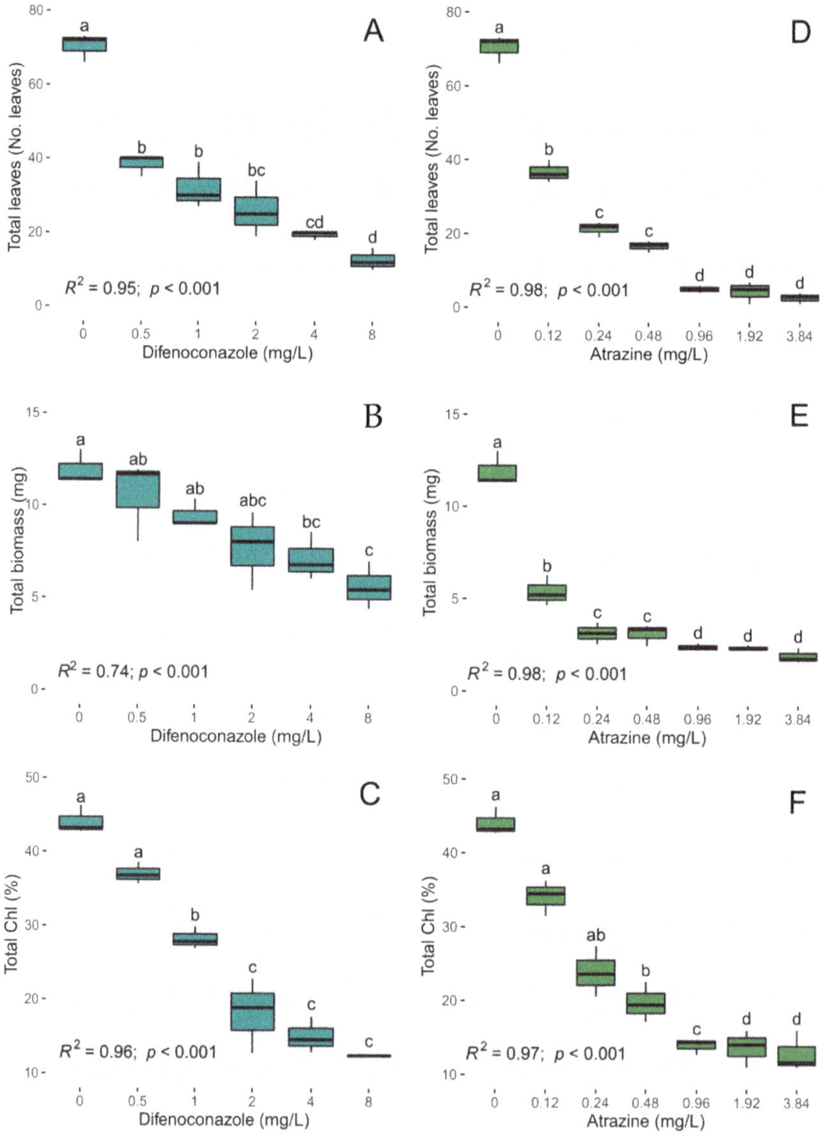

Figure 1. Effects of difenoconazole (**A–C**) and atrazine (**D–F**) after seven days of exposure on the number of leaves, total biomass, and total chlorophyll (Chl) content of *Lemma minor*. Different lowercase letters denote significant differences in the means at $p \leq 0.05$ (pairwise comparisons) between different pesticide concentrations.

Table 2. Summary of the GLM for the number of leaves, biomass, and total chlorophyll of *L. minor* and for the percentage of mortality of *D. magna* after being exposed to different concentrations (treatments) of difenoconazole and atrazine pesticides. Significant differences are indicated with the *p*-values in bold.

Species	Pesticide	Variable	Source of Variation	Df	*F*	*p*
Lemna minor	Difenoconazole	No. of leaves	Treatments	12	34.38	**<0.001**
		Biomass	Treatments	12	6.95	**0.003**
		Total chlorophyll	Treatments	12	70.64	**<0.001**
	Atrazine	No. of leaves	Treatments	12	67.94	**<0.001**
		Biomass	Treatments	12	103.83	**<0.001**
		Total chlorophyll	Treatments	12	72.20	**<0.001**
Daphnia magna	Difenoconazole	Mortality	Treatments	12	37.09	**0.001**
	Atrazine	Mortality	Treatments	12	11.92	**<0.001**

For atrazine, the highest number of leaves is observed in the control, which is different from the other concentrations, while there are no significant differences in the number of leaves between the concentrations of 0.24 and 0.48 mg/L, and the highest concentrations are the ones that have a negative effect on the number of leaves of *L. minor* (Figure 1). Regarding biomass, the highest biomass is observed in the control. There is a marked significant decrease towards higher concentrations of atrazine, but no significant differences are observed between these concentrations (Figure 1). The total content of chlorophyll is significantly higher in the control and in the concentration of 0.12 mg/L of atrazine, and it decreases towards the highest concentrations, with the concentrations of 0.92, 1.92, and 3.84 mg/L of atrazine being the doses with the greatest effect on the chlorophyll content of *L. minor*, although there are no significant differences between them (Figure 1).

In general, significant differences are found for the effects of different concentrations (control vs. highest concentration) of both pesticides and their photodegradation products on *L. minor*, but some exceptions are observed in the response of *L. minor* between the type of pesticide (e.g., Atrazine vs. A-Atrazine) (Table 3). For difenoconazole, in terms of number of leaves, biomass, and total chlorophyll content, both the pesticide and its two photodegradation products have the same effect (Figure 2).

Table 3. Summary of the GLM for the number of leaves, biomass, and total chlorophyll of *L. minor* and for the percentage of mortality of *D. magna* after being exposed to different concentrations (treatments) of pesticides and their photodegradation products (pesticides). Significant differences are indicated with the *p*-values in bold.

Species	Pesticide	Variable	Source of Variation	Df	*F*	*p*
L. minor	Difenoconazole	No. of leaves	Treatments	16	288.10	**<0.001**
			Pesticides	14	0.82	0.443
		Biomass	Treatments	16	25.02	**<0.001**
			Pesticides	14	0.13	0.874
		Total chlorophyll	Treatments	16	165.13	**<0.001**
			Pesticides	14	0.08	0.920
	Atrazine	No. of leaves	Treatments	10	104.38	**<0.001**
			Pesticides	9	41.89	**<0.001**
		Biomass	Treatments	10	78.27	**<0.001**
			Pesticides	9	6.40	**0.032**
		Total chlorophyll	Treatments	10	16.80	**0.003**
			Pesticides	9	5.07	0.051
D. magna	Difenoconazole	Mortality	Treatments	10	9.24	**0.009**
			Pesticides	9	6.60	**0.010**
	Atrazine	Mortality	Treatments	10	8.69	**0.016**
			Pesticides	9	4.67	0.060

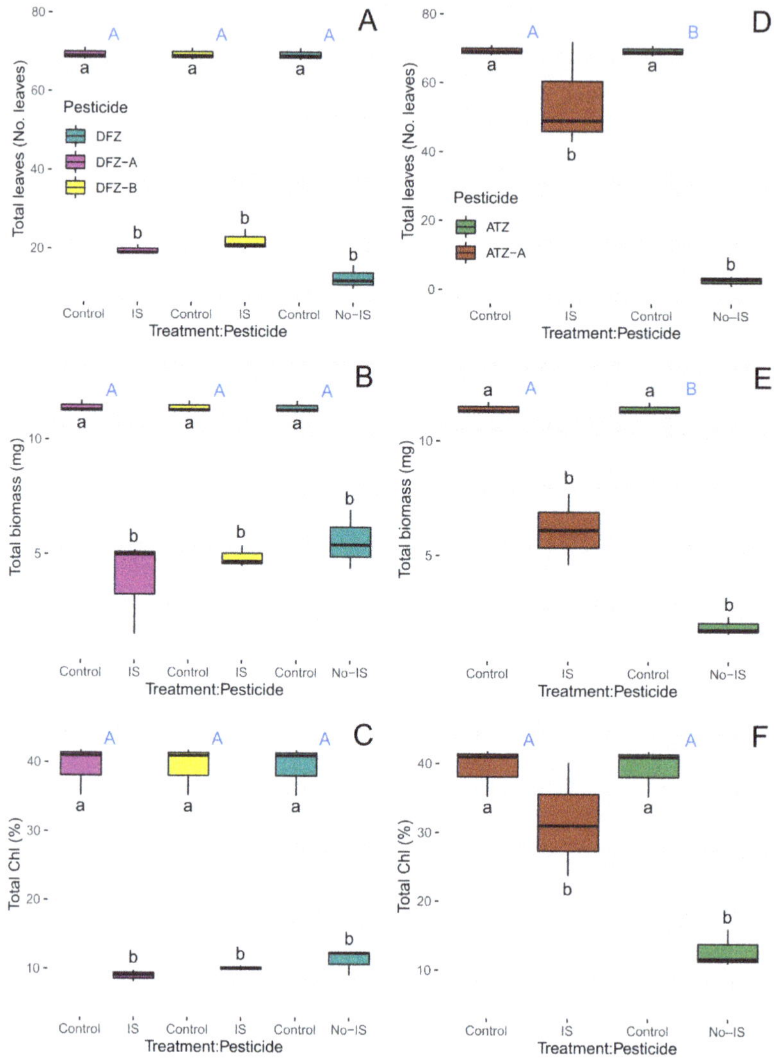

Figure 2. Effects of difenoconazole (**A–C**) and atrazine (**D–F**) and their photodegradation products after seven days of exposure on the number of leaves, biomass, and total chlorophyll (Chl) content of *L. minor*. Control refers to a concentration of 0 mg/L, whereas IS (irradiated solution) refers to highest concentration used in the bioassays, i.e., 8 mg/L for difenoconazole and its photodegradation products, and 3.84 mg/L for atrazine and its photodegradation product. Different letters denote significant differences in the means at $p \leq 0.05$ (pairwise comparisons) between different treatment (blue uppercase letters) and pesticide concentrations (lowercase letters).

For atrazine, there are significant differences in the effects of the pesticide and its photodegradation product on the number of leaves, biomass, and total chlorophyll content of *L. minor* (Figure 2).

3.3. Toxicity Tests with D. magna

Both difenoconazole and atrazine have a significant effect on *D. magna* neonates (Table 2). For difenoconazole, in the control, no mortality is observed in the *D. magna* neonates, while neonatal mortality is significantly higher at the concentrations of 0.8 and 1.6 mg/L of difenoconazole (Figure 3). For atrazine, no mortality of *D. magna* neonates is observed in the control nor in the concentrations of 10 and 20 mg/L. At 40, 60, and 80 mg/L concentrations of atrazine, a significant increase in the mortality of *D. magna* neonates is observed (Figure 3).

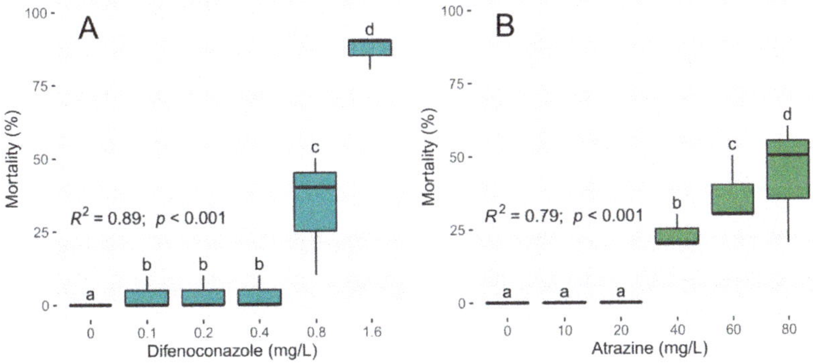

Figure 3. Effects of difenoconazole (**A**) and atrazine (**B**) after 48 h of exposure on the percentage of mortality of *Daphnia magna*. Different lowercase letters denote significant differences in the means at $p \leq 0.05$ (pairwise comparisons) between different pesticide concentrations.

In general, significant differences are found for the effects of different concentrations (control vs. highest concentration) of both pesticides and their photodegradation products on *D. magna*, but some exceptions are observed in the mortality response of *D. magna* neonates between the type of pesticide (e.g., Difenoconazole vs. A-Difenoconazole) (Table 3). For difenoconazole, there is a significant difference between the pesticide and its photodegradation products, with neonatal mortality being higher with the pesticide than with the two photodegradation products (Figure 4). In the case of atrazine, there is no significant difference in *D. magna* neonate mortality between the pesticide and its photodegradation product (Figure 4).

Regarding lethal concentrations of difenoconazole, the dose that is necessary to cause death in 50% of individuals is LC_{50}—48 h = 0.97 mg/L (0.85–1.08, 95% confidence interval). LC_{20}—48 h and LC_{10}—48 h values are determined for 0.66 mg/L (0.55–0.77) and 0.53 mg/L (0.41–0.65) of difenoconazole, respectively (Figure S1).

For atrazine, the LC_{50}—48 h value is 86.19 mg/L (66.04–106.34), being higher than the highest concentration evaluated in this study (80 mg/L). The dose to cause death in 20% and 10% of *D. magna* are LC_{20}—48 h = 40.63 mg/L (30.62–50.64) and LC_{10}—48 h = 26.17 mg/L (15.28–37.06) of atrazine, respectively (Figure S1).

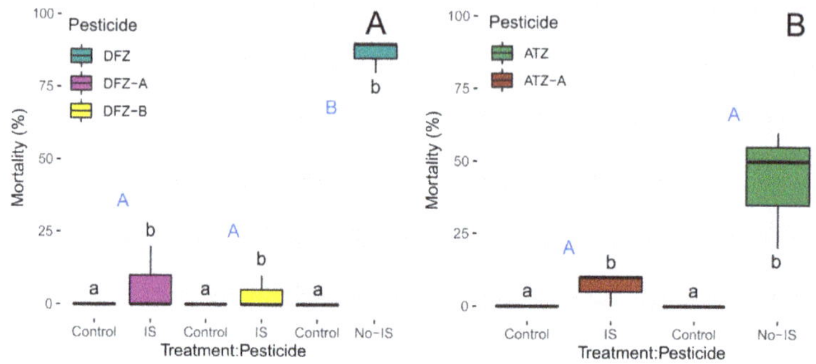

Figure 4. Effects of difenoconazole (**A**) and atrazine (**B**) and their photodegradation products after 48 h of exposure on the percentage of mortality of *D. magna*. Control refers to a concentration of 0 mg/L, whereas IS (irradiated solution) refers to highest concentration used in the bioassays, i.e., 1.6 mg/L for difenoconazole and its photodegradation products, and 80 mg/L for atrazine and its photodegradation product. Different letters denote significant differences in the means at $p \leq 0.05$ (pairwise comparisons) between different treatment (blue uppercase letters) and pesticide concentrations (lowercase letters).

4. Discussion

These results show how the two pesticides, difenoconazole and atrazine, and their photodegradation products cause a toxic effect on the growth of *L. minor* and on the mortality of *D. magna* neonates. We found that pesticides in their original state are a serious threat to aquatic biota and that photodegradation products remain toxic to aquatic organisms even though they are degraded. Our findings show that the number of leaves, biomass, and chlorophyll content of *L. minor*, as well as the mortality of *D. magna*, are sensitive metrics to quantify the toxic effects of pesticides and photodegradation products in aquatic environments. We believe that atrazine is a pesticide with a broad toxicological spectrum for plants and low toxicity for crustaceans at <20 mg/L concentrations in aquatic environments, whereas difenoconazole shows low toxicity for plants and high toxicity for crustaceans from 0.1 mg/l concentrations. Therefore, our results complement the information presented in other studies, where no effects of atrazine on *D. magna* was found [3,27,28]. In the research of Klematová et al. [3], they carried out a homogeneous photocatalytic degradation of atrazine through the photo-Fenton system and toxicity tests on *D. magna* and *L. minor*, showing that atrazine was toxic for *L. minor*, but it did not affect *D. magna*. Consequently, they established that photocatalytic degradation reduces the negative effect of atrazine on *D. magna*, while photodegradation products still negatively affect the growth of *L. minor*. Here, we demonstrated that when using a pesticide of different nature, aquatic biota may also response using different pathways.

In the case of difenoconazole, the work carried out by Man et al. [9] supports our results. They photodegraded difenoconazole in water as well as in soil, and they concluded that the toxicity caused in *D. magna* by the photodegradation products is significantly lower than the parent compound difenoconazole. Other researchers evaluated *D. magna* against difenoconazole and determined that there are effects on antioxidant and detoxifying enzymes, and on the lipid peroxidation of crustacean [4].

The fact that we could establish the toxicity that these pesticides may cause in aquatic biota provides us the context of what happens to these chemicals after they are applied to crops. It is known that the transformation of pesticides under different processes (photodegradation, hydrolysis, photolysis, physical-chemical conditions, degradation by microorganisms, etc.) produces new compounds that are released into the environment; however, we do not fully know the role these compounds play in the ecosystem.

The approach that we describe in this work allows us to evaluate the status of two common species that are part of the aquatic biota and that are known as indicators of ecotoxicity caused by pesticides that are used worldwide. Toxicological assays using standard indicators are useful, but complementary tests using autochthonous organisms would ideally improve the understanding of how pesticides affect local aquatic biota and provide better clues for regulating pesticide applications.

The concentrations of the pesticides and their degraded products assessed in this study may be used as reference information to establish potential hazardous effects in natural water bodies, as stated by Lamkhanter et al. [29]. To complement this study, we suggest performing assays to evaluate the combined effects of pesticides and their photodegradation products e.g., [3], or the combined effects of pesticides with other pollutants, such as microplastics. Additionally, the application of other bioindicators, such as fish, macroinvertebrates, cyanobacteria, or algae, is recommended in order to better understand how pesticides may affect natural ecosystems [30,31]. Another future approach can be the implementation of longer exposure times to toxicants (chronic toxicity) since pesticides can actually be present in the environment for very long periods, and organisms can experience chronic effects from exposure to pesticides and their degraded products [31].

Currently, pesticides and their degradation products have been identified in air, water, and soil in all geographic regions, including those that are very remote from the original site of their environmental release [32]. Knowing the agricultural practices in some countries, such as Ecuador, where banana and sugar cane plantations (crops commonly using difenoconazole and atrazine) were the crops with the largest irrigated area compared to the planted area in 2020, with 91.5% and 94.4%, respectively [17], displays a snapshot of current agricultural practices, where pesticides and their degradation products are potentially transported by irrigation systems to aquatic ecosystems, thereby contaminating water resources. For instance, Ochoa-Cueva et al. [20] reported that several sites in southern Ecuador present a high risk of pesticide exposure due to the indiscriminate application of pesticides across croplands. We recommend that in areas seriously exposed to pesticides, an irrigation water treatment based on photodegradation or advanced oxidation processes [33] may minimize the negative impacts of pesticides to the environment. However, as we found that degraded pesticide products can remain toxic in water, it is highly recommended the application of organic farming and the restriction of the use of pesticides, or only allowing those that degrade quickly and are less harmful to the ecosystem [34].

5. Conclusions

The results of the present investigation confirm that number of leaves, biomass, and chlorophyll content of *L. minor* and the mortality of *D. magna* are sensitive response variables that can be used to determine the toxicity of pesticides and photodegradation products in aquatic environments.

Pesticides, such as difenoconazole and atrazine, are a serious threat to aquatic biota and, after they are degraded by, for example, UV exposure, the resulting compounds remain toxic in the environment.

Supplementary Materials: The following supporting information can be downloaded at https://www.mdpi.com/article/10.3390/toxics11030213/s1, Table S1: Formulation for *Lemna minor* nutrient solution according to the Standard Methods 8211 [11]. A, B, and C refer to the stock solution to be prepared. Table S2: Formulation for preparing reconstituted freshwater for *Daphnia magna* according to the Standard Methods 8010:I [11]. Figure S1. Dose–response relationships between the percentage of mortality of *Daphnia magna* neonates and the concentrations of difenoconazole (**A**) and atrazine (**B**).

Author Contributions: Conceptualization, S.A.R.; methodology, S.A.R. and J.M.H.; formal analysis, J.M.H. and C.I.A.; writing—original draft preparation, J.M.H. and S.A.R.; writing—review and editing, C.I.A. and D.R.A.; funding acquisition, S.A.R. All authors have read and agreed to the published version of the manuscript.

Funding: This research was funded by the Universidad Técnica Particular de Loja through the Project: "EVALUATION OF PHOTOCATALYSTS FOR WATER DECONTAMINATION" Code: PROY_INNOV_QUI_2018_2455 Dimension: 212010014.

Data Availability Statement: The data presented in this study are available from the corresponding author upon request. The data are not publicly available due to further analyses and assays with the data used here.

Acknowledgments: We gratefully acknowledge Diego Ernesto Maza Estrada and Cristian David Jumbo Becerra for their assistance in laboratory analyses and David Emmanuel del Pozo Meza for cultivating *Lemna minor* individuals.

Conflicts of Interest: The authors declare no conflict of interest. The funders had no role in the design of the study; in the collection, analyses, or interpretation of data; in the writing of the manuscript; or in the decision to publish the results.

References

1. Ferhi, S.; Vieillard, J.; Garau, C.; Poultier, O.; Demey, L.; Beaulieu, R.; Penalva, P.; Gobert, V.; Portet-Koltalo, F. Pilot-scale direct UV-C photodegradation of pesticides in groundwater and recycled wastewater for agricultural use. *J. Environ. Chem. Eng.* **2021**, *9*, 106120. [CrossRef]
2. Severo, E.S.; Marins, A.T.; Cerezer, C.; Costa, D.; Nunes, M.; Prestes, O.D.; Zanella, R.; Loro, V.L. Ecological risk of pesticide contamination in a Brazilian river located near a rural area: A study of biomarkers using zebrafish embryos. *Ecotoxicol. Environ. Saf.* **2020**, *190*, 110071. [CrossRef] [PubMed]
3. Klementová, Š.; Hornychová, L.; Šorf, M.; Zemanová, J.; Kahoun, D. Toxicity of atrazine and the products of its homogeneous photocatalytic degradation on the aquatic organisms Lemna minor and Daphnia magna. *Environ. Sci. Pollut Res.* **2019**, *26*, 27259–27267. [CrossRef] [PubMed]
4. Moreira, R.; De Araujo, G.; Rego, A.; Daam, M.; Rocha, O.; Soares, A.; Loureiro, S. Effects of abamectin-based and difenoconazole-based formulations and their mixtures in Daphnia magna: A multiple endpoint approach. *Ecotoxicology* **2020**, *29*, 1486–1499. [CrossRef]
5. Inticher, J.J.; Cabrera, L.C.; Guimarães, R.E.; Zorzo, C.F.; Pellenz, L.; Seibert, D.; Borba, F.H. Advanced treatment of water contaminated with atrazine, difenoconazole and fipronil mixture, its by-products and bio-toxicity levels. *J. Environ. Chem. Eng.* **2021**, *9*, 105883. [CrossRef]
6. SGA. Sistema Globalmente Armonizado de Clasificación y Etiquetado de Productos Químicos. 2019. Available online: https://www.un-ilibrary.org/content/books/9789210040853 (accessed on 9 January 2023).
7. Tiwari, M.K.; Guha, S. Kinetics of biotransformation of chlorpyrifos in aqueous and soil slurry environments. *Water Res.* **2014**, *51*, 73–85. [CrossRef]
8. OMS. *Clasificación Recomendada por la OMS de los Plaguicidas por el Peligro que Presentan*; 2019; ISBN 9789240016057. Available online: https://cdn.who.int/media/docs/default-source/chemical-safety/pesticides/9789240016057-corrigenda-sp.pdf?sfvrsn=539aab18_17 (accessed on 9 January 2023).
9. Man, Y.; Stenrød, M.; Wu, C.; Almvik, M.; Holten, R.; Clarke, J.L.; Yuan, S.; Wu, X.; Xu, J.; Dong, F.; et al. Degradation of difenoconazole in water and soil: Kinetics, degradation pathways, transformation products identification and ecotoxicity assessment. *J. Hazard. Mater.* **2021**, *418*, 126303. [CrossRef]
10. Evgenidou, E.; Fytianos, K. Photodegradation of triazine herbicides in aqueous solutions and natural waters. *J. Agric. Food Chem.* **2002**, *50*, 6423–6427. [CrossRef]
11. Baird, R.; Bridgewater, L. *Standard Methods for the Examination of Water and Wastewater*, 23rd ed.; American Public Health Association: Washington, WA, USA, 2017; ISBN 978-087553-287-5.
12. González, C.; Vallarino, A. *Bioindicadores: Guardianes de Nuestro Futuro Ambiental*; González, C., Vallarino, A., Perez, J., Low Pfeng, A., Eds.; 2014; ISBN 9786078429059. Available online: https://agua.org.mx/wp-content/uploads/2017/11/Bioindicadores-Guardianes-de-nuestro-futuro-ambiental.pdf (accessed on 9 January 2023).
13. Sikorski, Ł.; Baciak, M.; Bęś, A.; Adomas, B. The effects of glyphosate-based herbicide formulations on Lemna minor, a non-target species. *Aquat. Toxicol.* **2019**, *209*, 70–80. [CrossRef]
14. Modlitbová, P.; Novotný, K.; Pořízka, P.; Klus, J.; Lubal, P.; Zlámalová-Gargošová, H.; Kaiser, J. Comparative investigation of toxicity and bioaccumulation of Cd-based quantum dots and Cd salt in freshwater plant *Lemna minor* L. *Ecotoxicol. Environ. Saf.* **2018**, *147*, 334–341. [CrossRef]
15. Núñez, M.; Hurtado, J. Bioensayos de toxicidad aguda utilizando Daphnia magna Straus (Cladocera, Daphniidae) desarrollada en medio de cultivo modificado. *Rev. Peru Biol.* **2005**, *12*, 165–170. [CrossRef]
16. Kojima, F.; Noldin, J.A.; Resgalla, C. Toxicidade aguda e análise de risco de herbicidas e inseticidas utilizados na lavoura do arroz irrigado sobre o cladócero Daphnia magna. *Capa* **2006**, *16*, 93–100.
17. INEC. Módulo de Información Ambiental y Tecnificación Agropecuaria. Available online: https://www.ecuadorencifras.gob.ec/documentos/web-inec/Encuestas_Ambientales/Modulo_Ambiental_ESPAC_2020/PRINC_RESUL_MOD_AGROTEC_2020_08_4.pdf (accessed on 9 January 2023).
18. FAO. FAOSTAT: Pesticides Indicators. Food and Agriculture Organization of the United Nations. Available online: https://www.fao.org/faostat/en/#data/EP/visualize (accessed on 9 January 2023).

19. Naranjo, A. *La Otra Guerra: La Situación de los Plaguicidas en el Ecuador*; Maldonado, A., Chérrez, C., Bravo, E., Eds.; 2017. Available online: https://www.accionecologica.org/la-otra-guerra-situacion-de-los-plaguicidas-en-ecuador/ (accessed on 9 January 2023).
20. Ochoa-Cueva, P.A.; Arteaga, J.; Arévalo, A.P.; Kolok, A.S. A potential pesticides exposure index (PPEI) for developing countries: Applied in a transboundary basin. *Integr. Environ. Assess. Manag.* **2021**, *2021*, 235267474. Available online: https://www.semanticscholar.org/paper/A-potential-pesticides-exposure-index- (accessed on 9 January 2023).
21. Arnon, D.I. Copper Enzymes in Isolated Chloroplasts. Polyphenoloxidase in Beta Vulgaris. *Plant Physiol.* **1949**, *24*, 1–15. [CrossRef]
22. Marr, I.L.; Suryana, N.; Lukulay, P.; Marr, M.I. Determination of chlorophyll a and b by simultaneous multi-component spectrophotometry. *Fresenius J. Anal. Chem.* **1995**, *352*, 456–460. [CrossRef]
23. R Foundation for Statistical Computing. *A Language and Environment for Statistical Computing*; R Foundation for Statistical Computing: Vienna, Austria, 2022.
24. Russell, V.L.; Buerkner, P.; Giné-Vázquez, I.; Herve, M.; Jung, M.; Love, J.; Miguez, F.; Riebl, H.; Singmann, H. emmeans: Estimated Marginal Means, aka Least-Squares Means. R package version 1.8.4-1 2023. Available online: https://cran.r-project.org/web//packages/emmeans/emmeans.pdf (accessed on 9 January 2023).
25. Ritz, C.; Strebig, J.C. Drc: Analysis of Dose-Response Curves. 2016. Available online: https://mran.microsoft.com/snapshot/2023-01-28/web/packages/drc/drc.pdf (accessed on 9 January 2023).
26. Tagun, R.; Boxall, A.B.A. The Response of Lemna minor to Mixtures of Pesticides That Are Commonly Used in Thailand. *Bull. Environ. Contam. Toxicol.* **2018**, *100*, 516–523. [CrossRef] [PubMed]
27. Palma, P.; Palma, V.L.; Fernandes, R.M.; Soares, A.M.V.M.; Barbosa, I.R. Acute toxicity of atrazine, endosulfan sulphate and chlorpyrifos to Vibrio fischeri, Thamnocephalus platyurus and Daphnia magna, relative to their concentrations in surface waters from the Alentejo region of Portugal. *Bull. Environ. Contam. Toxicol.* **2008**, *81*, 485–489. [CrossRef] [PubMed]
28. Kelly, M.R.; Cohen, R.A. The Effects of an Herbicide and Antibiotic Mixture on Aquatic Primary Producers and Grazers. *Bull. Environ. Contam. Toxicol.* **2018**, *101*, 556–561. [CrossRef]
29. Lamkhanter, H.; Frindy, S.; Park, Y.; Sillanpää, M.; Mountacer, H. Photocatalytic degradation of fungicide difenoconazole via photo-Fento process using α-Fe2O3. *Mater. Chem. Phys.* **2021**, *267*, 124713. [CrossRef]
30. Kafula, Y.A.; Thoré, E.S.J.; Philippe, C.; Munishi, L.K.; Moyo, F.; Vanschoenwinkel, B.; Brendonck, L. Environmental risks of a commonly used pyrethroid: Insights from temporary pond species of the Lake Manyara Basin, Tanzania. *Sci. Total Environ.* **2023**, *868*, 161698. [CrossRef]
31. Thoré, E.S.J.; Philippe, C.; Brendonck, L.; Pinceel, T. Towards improved fish tests in ecotoxicology—Efficient chronic and multi-generational testing with the killifish Nothobranchius furzeri. *Chemosphere* **2021**, *273*, 129697. [CrossRef]
32. Cruz, G.; Julcour, C.; Jáuregui-Haza, U. El Estado actual y perspectivas de la degradación de pesticidas por procesos avanzados de oxidación. *Rev. Cubana De Química* **2017**, *29*, 492–516. Available online: scielo.sld.cu/scielo.php?script=sci_arttext&pid=S2224-54212017000300013 (accessed on 9 January 2023).
33. Aguilar, S.D.; Ramos, D.R.; Santaballa, J.A.; Canle, M. Preparation, characterization and testing of a bulky non-supported photocatalyst for water pollution abatement. *Catal. Today* **2022**. [CrossRef]
34. Almeida, R.A.; Lemmens, P.; Cours, M.; Denys, L.; Adriaens, D.; Packet, J.; Vondorickx, J.; Vercauteren, T.; Parmentier, K.; Knockaert, M.; et al. A moderate differential effect of organic and conventional agriculture across taxonomic groups inhabiting farmland ponds. *Freshw. Biol.* **2023**, *2023*, 1–14. Available online: https://pureportal.inbo.be/en/publications/a-moderate-differential-effect-of-organic-and-conventional-agricu (accessed on 9 January 2023). [CrossRef]

toxics

MDPI

Article

Experiments on Pilot-Scale Constructed Floating Wetlands Efficiency in Removing Agrochemicals

George Pavlidis [1], Ioanna Zotou [1], Helen Karasali [2], Anna Marousopoulou [2], Georgios Bariamis [3], Ioannis Nalbantis [1] and Vassilios A. Tsihrintzis [1,*]

[1] Centre for the Assessment of Natural Hazards and Proactive Planning & Laboratory of Reclamation Works and Water Resources Management, School of Rural, Surveying and Geoinformatics Engineering, National Technical University of Athens, 9 Heroon Polytechniou St., Zographou, 15780 Athens, Greece
[2] Laboratory of Chemical Control of Pesticides, Department of Pesticides Control and Phytopharmacy, Benaki Phytopathological Institute, 8 Stef. Delta St., Kifissia, 14561 Athens, Greece
[3] Department of Water Resources and Environmental Engineering, School of Civil Engineering, National Technical University of Athens, Greece, 9 Heroon Polytechniou St., Zographou, 15780 Athens, Greece
* Correspondence: tsihrin@survey.ntua.gr or tsihrin@otenet.gr; Tel.: +30-(21)-07722620

Abstract: The efficiency of constructed floating wetlands (CFWs) in their ability to remove agrochemicals (nutrients and pesticides) is here investigated in a series of pilot-scale systems. Four experimental CFWs were designed and constructed; three of them were planted with the aquatic plant species *Lemna minor*, *Azolla pinnata* and *Eichhornia crassipes*. The fourth did not contain any plants and was used as the control. The aim of the study was to evaluate the efficiency of CFW containing aquatic macrophytes in the reduction of pesticides and nutrients, under field conditions. The CFWs operated continuously from May 2021 to September 2021, and their removal efficiencies of nitrogen and phosphorus ions, and five commonly used pesticides were examined. The CFW systems were fed daily with agricultural wastewater which was prepared by mixing a fertilizer and predetermined doses of pesticides. The hydraulic residence time was kept at 14 days. Samples were collected on a weekly basis from both the influent and the effluent of each experimental tank, and were subsequently analyzed in the laboratory. HPLC-DAD and Ion Chromatography were implemented for sample analysis following a very simple sample preparation. Reductions for nutrient ranged from no reduction to 100% removal, whereas for pesticides these varied from no reduction to 98.8% removal, indicating that these systems can be used as efficient and low-cost pollution control technologies for agrochemical wastewater treatment. Significant reduction for certain pesticides was also observed in the algae control tank, thus, proving the efficiency of algae in organic pollution reduction, and recognizing the limitations of aquatic plant use in decontamination.

Keywords: constructed floating wetlands; agricultural wastewater treatment; nitrogen; phosphorus; pesticides; removal

1. Introduction

During the 20th and 21st century, natural water bodies have been subjected to severe environmental pressure that resulted from both natural and anthropogenic causes [1]. The major reasons for ecosystem impairment are population growth, urbanization and intensification of industry and agriculture [2]. The use of agrochemicals plays a significant role in agricultural non-point source pollution of water bodies, posing a serious danger to drinking water resources and aquatic ecosystems. Therefore, agrochemical pollution has become a key issue of concern for the scientific community, NGOs and governments worldwide. To abate pollution, proper on-site management measures and/or use of simple and low-cost treatment technologies are required [3]. Conventional techniques designed for the treatment of wastewaters (e.g., ion exchange, adsorption, reverse osmosis, chemical precipitation, electrochemical treatment) often fail to completely remove various kinds of water contaminants; furthermore, they are expensive, energy-intensive and non-eco-friendly. An increasing need thus arises for adopting cost-effective and environmentally-friendly purification technologies for removing pollutants from water and for restoring aquatic ecosystems [4–7]. Phytoremediation is such a technology, which makes use of aquatic plants and the potential of the latter to absorb and accumulate nutrients or other substances in their tissues, thus remediating wastewater [8]. Through this process, organic and inorganic pollutants are transferred through the root system to the upper part of the plant, thus naturally purifying the contaminated soil or water [9].

Several techniques for mitigation of agricultural pollution have been proposed, including, among others, natural pollution abatement systems, such as constructed wetlands, stabilization ponds, algae systems, and agroforestry systems [10,11]. Constructed wetlands (CWs) are natural treatment systems wherein remediation of contaminated water is implemented through the physical, chemical and biological processes naturally occurring during the interaction of water and/or soil, plants and microorganisms [12]. The effectiveness of the aforementioned systems, especially when combined with aquatic macrophytes, has been well-documented for different kinds of wastewater, such as industrial, municipal, agricultural, mine waste and stormwater [5,13–20]. Various system types can be used in the treatment of the vast majority of pollutants, including organic substances, fecal pollutants, metals, nitrogen and phosphorus, pesticides in all forms, as well as PAHs, PCBs, PCEs, BTEX compounds and hydrocarbons [11,21–31]. A great variety of polluting substances, such as nutrients, pesticides, heavy metals, sediments and bacterial contaminants have been investigated and proven to be efficiently removed or diminished by employing aquatic macrophytes in CWs [32]. However, the effectiveness of natural pollution control systems on pesticide treatment has not been studied adequately, especially with regard to the water-based systems.

Aquatic plants established in CWs may also favor water transparency by reducing water velocity and the resultant high concentration of suspended solids in the system [20]. Although in most studies the use of CWs has had an overall beneficial effect in remediating effluents from agricultural activity, results in terms of pollutant removal effectiveness greatly vary as a result of differing environmental and design factors [33], such as climate, hydraulic loading rate, pollutant loading rate, hydraulic retention time, growth media, water depth, vegetation type and age, and percent of vegetation coverage [32]. Díaz et al. [32] reported that CWs performance in removing nitrate from agricultural runoff varies from "negative", namely being itself a nitrate source, to 98%. Regarding total phosphorus, CWs efficiency has been found to range between non-significant to as high as 80%, whereas pesticides can be removed from agricultural effluents by 0 to 100%.

Aquatic macrophytes being exploited in CWs are subdivided into three general categories, that is, free-floating (e.g., *Pistia stratiotes, Azolla pinnata, Eichhornia crassipes, Lemna* spp.), submerged (e.g., *Myriophyllum aquaticum*) and emergent (e.g., *Typha* spp. and *Phragmites*), with plant species of the third type being the most commonly utilized in phytoremediation applications, due to their greater availability, high growth rate and ease in terms of harvesting and stocking [5,8,34]. Constructed floating wetlands (CFWs), also referred to as artificial floating islands (AFI), are a variant of CWs employing floating vegetation to remediate water [7].

As regards the aquatic plants used in the present study, *L. minor* is a free-floating aquatic species, commonly investigated for its potential in assimilating nutrients (nitrogen and phosphorus) from various sources effluents [2]. Priya et al. [35] evaluated the efficiency of *Lemna minor* in treating organic waste and removing nutrients from domestic wastewater. The experiment was conducted in a pilot scale apparatus, after primary and secondary treatment of the wastewater. The results showed good performance of the aquatic plant, which reduced BOD and orthophosphate concentrations by 94.45% and 79.39%, respectively. Aziz et al. [36] comparatively assessed the performance of four different aquatic plants, among which was *Lemna minor*, and found that the latter was capable of reducing ammoniacal nitrogen (NH_4^+-N) from sewage by 80.4% after eight days of treatment. Sarkheil and Safari [37] demonstrated the potential of using duckweed in aquaculture as well. In their experiment, undertaken in a recirculating water system used to culture African cichlid, *L. minor* reduced the concentrations in total nitrogen ammonia and total phosphorus by 43.7% and 52.38%, respectively [37]. Liu et al. [38] investigated the efficiency of *L. minor* in removing nitrogen (N) and phosphorus (P) from industrial wastewater, in relation to the salinity level in water. Their findings showed that high salt stress combined with long periods of exposure inhibited duckweed capacity to absorb N and P, while for NaCl concentrations above 100 mM the aquatic plant had negative removal efficiency, functioning as a sink of N and P. Ceschin et al. [39] examined the phytoremediation performance of Lemna in a three-pool CW designed for treating municipal wastewater produced from the town of Forano in Central Italy. The results highlighted the adverse effect of Lemna overgrowth on treatment system efficiency. The authors stated that the successful application of duckweed in phytoremediation requires periodic harvesting, in order to avoid the development of an extended and thick mat, which in turn impedes light penetration and favors anaerobic conditions. Kamyab et al. [40] attempted to evaluate the capacity of duckweed and microalgae to remove nutrients from palm oil mill effluents as well as to further utilize the aquatic plants for fertilizer production. The removal rate achieved from the combined use of the two species attained 12.5%, 11.3% and 70.5% for NO_3^--N, NH_4^+-N and PO_4^{3-}-P [40]. In a similar experiment, Sudiarto et al. [41] compared Lemna with three additional aquatic plant species in terms of their ability to remove nutrients from treated livestock wastewater. The results demonstrated that duckweed was the most effective among the examined species in phosphorus uptake, achieving a removal rate of 36.15%. However, it was found to be unsuitable for nitrogen removal because of its low growth rate. Duckweed has also been successfully applied for the uptake of pesticides from agricultural runoff. Dosnon–Olette et al. [42], in a four-day experiment, tested the efficiency of Lemna along with four additional macrophyte species in removing dimethomorph and pyrimethanil, focusing also on the toxicity these substances exert on the aquatic plants depending on concentration levels. *Lemna minor* along with *S. polyrhiza* were found to be the most effective in fungicide removal with the former yielding a removal rate equal to 12% and 17% for pyrimethanil and dimethomorph, respectively. The authors also suggested that a concentration of 600 $\mu g\ L^{-1}$ for these fungicides is suitable, so as to not inhibit photosynthetic activity of the utilized macrophytes. In a similar study conducted by Dosnon–Olette et al. [43], *L. minor* exhibited again the highest performance among the examined species in uptaking copper sulphate, flazasulfuron and dimethomorph. In this 7-day experiment, Lemna achieved a removal rate of 50%, 11.5% and 42% for copper sulphate, dimethomorph and flazasulfuron, respectively. A concentration of 40 $\mu g\ L^{-1}$ for copper and fluzasulfuron and 400 $\mu g\ L^{-1}$ for dimethomorph were defined as the optimum ones for evaluating remediation efficiency of the examined plants based on the toxicity test undertaken. Dosnon–Olette et al. [44] also tested *L. minor* against *S. polyrhiza*, in terms of their removal efficiency and toxicity, solely considering dimethomorph. Lemna outgrew *S. polyhriza*, reaching a removal rate of 41 $\mu g\ g^{-1}$ of dimethomorph, for a 600 $\mu g\ L^{-1}$ concentration after 4 days of exposure. The authors also reported a strong positive relationship between initial population density and dimethomorph toxicity for both species. The pesticide removal efficiency of *L. minor* has also been examined against chlorpyrifos by Prasertsup and Ariyakanon [45]. The experiment was undertaken under laboratory greenhouse conditions and the results showed a considerable inhibition of the relative growth rate of the aquatic plant for chlorpyrifos concentration as high as 1000 $\mu g\ L^{-1}$, whereas the maximum removal yield was observed for 500 $\mu g\ L^{-1}$ of the examined insecticide, reaching 87% [45]. A lower removal yield was determined for two other herbicides, namely isoproturon and glyphosate, according to an experiment conducted by Dosnon–Olette et al. [46], where *Lemna minor* was able to uptake 25% and 8% of each of the two agrochemicals, respectively, after a 4-day exposure. Furthermore, several studies have examined duckweed response in terms of removal rate and sensitivity for a variety of additional pesticides (either alone or in mixture), including metolachlor, atrazine, metribuzin, lactofen, linuron, monolinuron, diuron, 2,4-D, alachlor, paraquat, propanil, among others [47–52].

Eichhornia crassipes, commonly known as water hyacinth (WH), which was also used in the present study experiments, is another example of free-floating macrophyte, well-known for its pollutant removal capacity. Its effectiveness in wastewater treatment, mainly due to the assimilation of nutrients and heavy metals, has been proved by several studies [8,16]. Fox et al. [53] examined WH efficiency in removing nitrogen under different concentrations ranging between 0 and 300 ppm. The results demonstrated a 60–85% nitrogen assimilation rate after a 4-week period, whereas it was also found that biomass production, although having a positive relationship with applied nitrogen rates, stops increasing for N concentrations above 80 ppm [53]. Nabi et al. [54] elaborated on a 30-day experiment to test WH nutrient performance for domestic wastewater. The results indicated that the aquatic plant was capable of removing 63.28% and 58.54% of Total Nitrogen (TN) and phosphorus, respectively. Osti et al. [55] employed WH to improve water quality in tilapia fishponds by reducing nitrogen and phosphorus. The results revealed a reduction in TN, Total Inorganic Nitrogen (TIN), Total Phosphorus (TP) and PO_4^{3-}-P, induced by WH, equal to 66%, 82%, 27% and 33%, respectively. A 24-day experiment was performed by Kutty et al. [56] in order to assess *Eichhornia crassipes* nutrient accumulation capacity from sewage treatment plant effluent. The aquatic macrophyte exhibited a removal yield of 81%, 67% and 92% for NH_3-N, P and NO_3^--N, respectively, whereas it also presented a considerable growth rate from the sixth day and until the end of the experiment. Additional studies having investigated removal efficiency and/or sensitivity of WH in nutrient-rich wastewater are those of Sooknah and Wilkie (2004), Chen et al. (2010), Aremu et al. (2012), Zhao et al. (2012), Wang et al. (2013) and Lima et al. (2018) [57–62], among others. Despite the great availability of experiments undertaken in order to assess nutrient removal efficiency of WH, only a few have investigated WH potential for removing pesticides from agricultural wastewater. Xia and Ma [63] examined ethion uptake capacity of WH by employing four different culture solutions, that is, non-sterile planted, sterile planted, non-sterile unplanted and sterile unplanted treatment. They found that WH accounted for 69% of the total removal of the utilized ethion after 240 h of incubation against 12% of removal caused by bacterial degradation [63], with the removal rate constant of ethion due to the aquatic plant estimated at 0.00730 h^{-1}. The removal capacity of *Eichhornia crassipes* against the organophosphate insecticide chlorpyrifos was investigated by Anudechakul at al. [64]. WH, along with the synergistic action of the bacterium *Acinetobacter* sp. strain WHA, achieved a removal rate constant 3.89–4.87 times higher (depending on chlorpyrifos applied concentrations) than that which occurred in the absence of plants [64]. Alencar et al. [65] comparatively examined *Pistia stratiotes* and water hyacinth regarding their efficiency in uptaking clomazone. WH was found to be more resistant in the presence of the examined herbicide, achieving, however, a lower removal yield compared to *P. stratiotes* (90 and 99.9% for WH and *P. stratiotes*, respectively) [65].

Finally, *Azolla pinnata* that was also examined in the present study, was previously investigated for its efficiency in treating four different wastewaters (domestic, municipal aquaculture and industrial) for agricultural re-use, where it recorded a complete removal of phosphorus and nitrogen compounds for all waste types tested, except for the municipal where a maximum of 75.7% was obtained for phosphorus [66]. Following the same rationale, Soman et al. [67] examined the ability of Azolla plants to remove nutrients from secondary treated wastewater and observed removals of 54.8% for ammonia, 50% for organic carbon, 71.4% for nitrites, 80.5% for total phosphorus, 91.7% for BOD and 87.4% for COD. The above findings were also supported by the study of Muvea et al. [68], as from their analysis it was observed that Azolla plants in a CFW with 10–14 days retention time efficiently removed from wastewater approx. 75% of nitrites, 32% of nitrates, 17% of ammonium and 50% of total phosphorus. However, these authors pointed out that a longer retention time would improve reductions. Another species of Azolla, the *Azolla filiculoides*, has also presented nitrogen, phosphorus and COD removals from secondary effluents treated for

28 days of up to 36%, 44% and 98.8%, respectively, thus posing that Azolla may be one of the most promising floating plants for CFW [69]. Finally, Akinbile et al. [66] reported metal removal in addition to nutrients, solids and turbidity, with removal rates reaching 70% for zinc, 99.6% for iron and 64% for magnesium.

The literature survey undertaken revealed that some work had been carried out on the potential of *Lemna minor*, *Azolla pinnata* and *Eichhornia crassipes* for removing nutrients from wastewater. Fewer studies, though, have comparatively examined the capacity of these species regarding pesticides uptake. Besides, to the best of our knowledge, there is no study having investigated the effectiveness of these macrophytes against a combination of nutrients and pesticides, which often co-exist in agricultural runoff. Therefore, the present study seeks to comparatively assess the removal efficiency of *Lemna minor*, *Azolla pinnata* and *Eichhornia crassipes*, at field-like conditions and pilot-scale systems, considering ammonium, phosphate, and nitrate ions from fertilizers, as well as the following five pesticides: (a) imidacloprid; (b) thiacloprid; (c) dimethomorph, (d) myclobutanil; and (e) difenoconazole. In order to quantify the removal rates achieved by CFW systems, the three plant species were established in respective pilot-scale CFWs continuously operating for a 16-week period, that is, from May to September 2021, whereas a fourth no-plant system, acting as control, was also established to test agrochemical removal in the absence of plants, induced though by other factors, such as hydrolysis, photolysis, bacterial and/or naturally developed algae degradation. The ultimate goal of the present experiment was to propose CFW systems for treating agricultural runoffs as well as fertilizer and pesticide residues from spraying equipment tanks, instead of discharging them, untreated, to surface water bodies. The novelty of the study lies on the fact that it provides information regarding pesticide degradation in CFW systems, where the current available literature is scarce. The paper provides new experimental data on the design and operation of these systems by examining three different aquatic macrophytes and a control algal system in parallel experiments under the same climatic conditions. A comparison is also made between warm and cold seasons accounting for the effect of temperature. Additionally, the experiment is conducted under actual field conditions in pilot-scale systems located in the open air and not in a closed laboratory. Thus, the effect of meteorological parameters can be fully examined.

2. Materials and Methods

2.1. Experimental Parameters

The study was conducted at the premises of the School of Rural Surveying and Geoinformatics Engineering at the National Technical University of Athens Campus located in Zographou, Attica, Greece (coordinates: 37°58′30.9″ N; 23°46′47.2″ E). The tanks used for the experiment were made of hard PVC plastic, with dimensions 1.40 × 0.75 m, and were placed on the roof of the building (Figure 1). The sides of the tank were covered by black plastic material to avoid below-water sunlight impact. The study was initiated on 1 May 2021 with the necessary plant growth procedures. Wastewater loading began on 14 May and continued until 3 September 2021. The artificial waste was prepared daily right before feeding, using tap water, commercial pesticides (Plant Protection Products-PPPs) and a water-soluble 31-11-11+TE (N-P-K+micronutrients) granular fertilizer. The exact quantities of the artificial wastewater input per tank were: for imidacloprid (20% *w/v*) 0.67 mL, for thiacloprid (24% *w/v*) 0.56 mL, for myclobutanil (12% *w/v*) 1.11 mL, for difenoconazole (25% *w/v*) 0.53 mL, for dimethomorph (50% *w/v*) 0.27 mL and for the fertilizer 6.45 g, thus reaching an influent pollutant concentration of 2 mg/L for the pesticides, 30 mg/L for TN and 10.65 mg/L for TP.

Influent and effluent samples were collected on a weekly basis to determine their concentrations and estimate pollutant reductions after treatment in the systems. To eliminate any degradation, samples were deep-frozen and stored in the dark until the time of sample preparation and analysis. Wastewater loading began 2 weeks after the installation of aquatic plants to allow for aquatic plant growth and multiplication, whereas the first sampling was done one week later. The water depth in each tank was maintained at about 30 cm using an overflow weir at the tank outlet (Figure 1e) and the volume for each tank was 305 L. Wastewater feeding was performed once a day for 4 days per week at equal doses using an inverse T-shaped PVC pipe with 16 2-mm holes in its horizontal arms that was installed at the inflow side of each tank [2]. The inlet wastewater volume loaded each time was 38.1 L, thus reaching a hydraulic retention time (HRT) of 14 days, also initially implemented for plant equilibration. Apart from the chemical parameters examined, several water quality parameters were also monitored (temperature, pH, Electrical Conductivity, Total Dissolved Solids, Salinity and water depth). Finally, a meteorological station positioned next to the experimental tanks recorded basic meteorological parameters (i.e., air temperature, relative air humidity, rainfall depth, wind speed and solar radiation) at an hourly time-step with the purpose of checking possible correlations between pollutant removal efficiency and meteorological conditions.

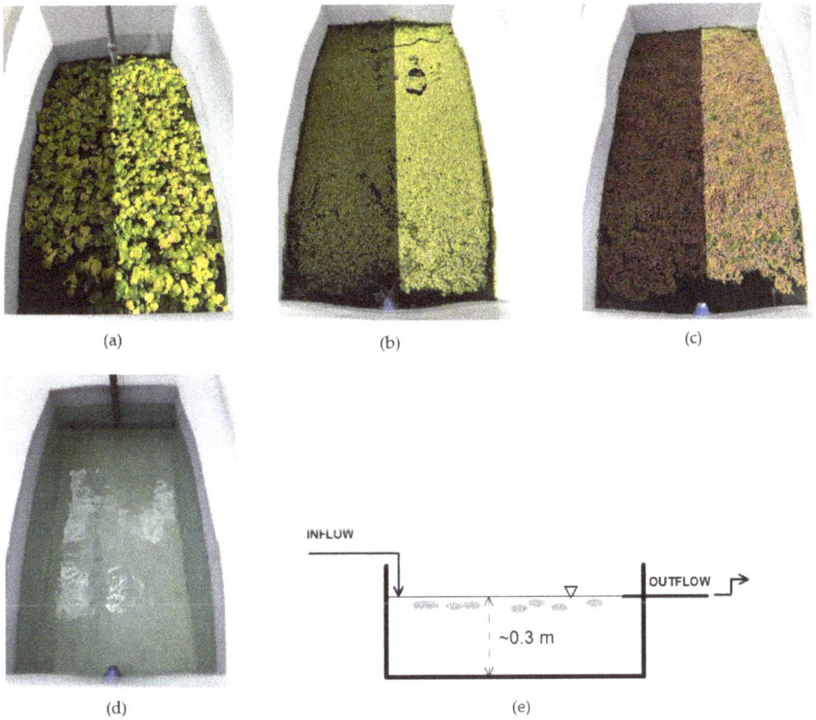

Figure 1. View of the main components of the experiment: (**a**) water hyacinth tank; (**b**) *Lemna minor* tank; (**c**) azola tank (**d**) control tank; (**e**) schematic representation of tanks.

In total, four similar tanks were used for the present experiment (Figure 1). The *L. minor* and *A. pinnata* plants used were obtained from a local nursery (Attica, Greece) and reproduced in the actual test tanks, and *Eichhornia crassipes* was available from previous experiments and was reproduced in order to be used for the present study. In parallel to the plant-containing tanks, a control tank without plants was run (however, algae developed with time, acting as an oxidation pond) and was treated the exact same way as the macrophyte-containing tanks.

Algae were eventually developed in the other three experimental tanks; therefore, the comparison with the control tank to see the effects of the two plant species was possible.

The active ingredients selected were among the most commonly used in the EU region and in Greece, and are important for agricultural production, whilst they present significantly different physicochemical and environmental fate properties (Table 1).

Table 1. The examined pesticides and their environmental fate endpoints [70–74].

Substance	NOEC (Studies with *L. gibba*)	Water Solubility (mg/L)	Koc (Adsorption) (mL/g)	DT50soil	Photolysis	Hydrolysis	DT50water	DT50 System	Type
Myclobutanil	105 mg as/L	132	225–920	191–1216 d lab/ 9–58 d field	Stable	Stable	4–20 d	415–838 d	fungicide
Imidacloprid	no data	600	109–411	27–180 d field	Yes DT50 = 1 h	Stable	>30 d	129 d	insecticide
Difenoconazole	no data	15	400–7730	20–242 d field	Stable	Stable	1–2 d	307–324 d	fungicide
Thiacloprid	46.8 mg as/L	184	393–870	6–16.8 d field	Stable (79.7 d)	Stable	n.a.	12.1–18 d	insecticide
Dimethomorph	no data	10.7–47.2	290–566	34–53.4 d lab/ 10–61 d field	Stable (86–107 d)	Stable	5–15 d	16–59 d	fungicide

2.2. Sample Preparation

Nutrient residues sample preparation included homogeneous mixing of the sample in a table shaker at 160 rpm and filtration via a 0.45 μm and a 0.22 μm syringe filter. Appropriate dilution (1:10) was performed for the inlet samples due to their high concentration for the IC technique.

For pesticide residue determination, a filtration through 0.45 μm GF-PET filter was applied before analytical determination.

2.3. Chemicals, Analyses and Instrumentation

Certified stock solutions at a concentration of 1000 mg L^{-1}, purchased from Dionex, USA, were used for nutrient determination after the appropriate dilution per ion in 18.3 MΩ ultrapure water. The mobile phases used were methanesulfonic acid and sodium carbonate-bicarbonate, also purchased from Dionex (Sunnyvale, CA, USA) and prepared in 18.3 MΩ ultrapure water. The analytical instrument used was a Dionex ICS-3000 (USA) system with IonPac AS 23 (4 × 250 mm) and CS16 (5 × 250 mm) columns with the respective column guards (Thermo, Waltham, MA, USA). Data acquisition and processing were performed using Chromeleon ver. 7 software. The method was validated and the recoveries for all analytes ranged between 84–111%, and the linearity coefficient values (r^2) achieved from seven points were higher than 0.995, whilst the Limit of Quantification (LOQ) was dependent on the ion, that is, 0.1 mg/L for PO_4^{3-}-P, 0.05 mg/L for NH_4^+-N, and 5 mg/L for NO_3^--N and was set as the lowest point concentration at the calibration curve.

Similarly, pesticide residues were determined using reversed-phase high-performance liquid chromatography with the diode array detection (HPLC-DAD) technique. A Nucleodur C-18 gravity 150 × 4.6 mm (5 μm) column was used for the quantitative determination of pesticide residues. Analytical standards of high purity were used for establishing linearity and linear range, and repeatability and accuracy of the analytical method. The analytical standard of imidacloprid (98.8%) was obtained from Bayer Crop Science, thiacloprid (98.8%) was obtained from Bayer Crop Science, dimethomorph (97.6%) was obtained from BASF, whereas myclobutanil (98.6%) and difenoconazole (99.9%) were obtained from Sigma Aldrich. The individual standard stock solutions were prepared after an appropriate dilution of the respective analytical standard to a final concentration of approximately 1 mg/mL.

Standard working solutions and their mixtures were prepared by independent dilutions of the stock solutions in acetonitrile. Acetonitrile (Merck, Darmstadt, Germany) and water (Fisher Scientific UK Limited, Loughborough, UK) were of HPLC grade. The mobile phase used for pesticide residues determination was ACN/0.1% Acetic acid aqueous solution 60/40 (*v/v*). Working standard solutions were freshly prepared from the individual stock solutions.

HPLC analysis was carried out using a Shimadzu UFLC instrument (Shimadzu, Japan), equipped with a diode array detection system (SPD-M20A), a column oven (CTO-20A), a degasser (DGU-20AS) and an autosampler (SIL-20AC). The substance-specific chromatographic parameters for the examined substances are presented in Table 2. Instrument control and post-run data treatment were performed using Shimadzu Lab Solution software, version 1.25. A representative chromatogram is presented in Figure 2, whereas chromatographical method parameters are presented in Table 2. The applied methods were fully validated with respect to linearity, specificity, accuracy (in terms of recovery), and limit of detection and quantification (LOD and LOQ, respectively). The LOQ was 0.01 μg mL^{-1}, for all analytes. Linearity of the chromatographic system was established using 5 calibration solutions in the range of 1 μg mL^{-1} to 50 μg mL^{-1} for the examined compounds. The correlation coefficient, as determined from the calibration curve, was 0.999 for all compounds. Recoveries ranged from 90 to 110% in all cases. Influent and effluent wastewater physicochemical parameters were measured using a portable YSI (Yellow Springs, OH, USA) Pro Plus multimeter.

Table 2. Chromatographical parameters.

Analyte	Detector/ Wavelength	Flow Rate (mL/min)	Injection Volume (μL)	Retention Time (min)	Column Oven Temperature (°C)
Imidacloprid	UV (230 nm)	1.0	10	1.85	35
Thiacloprid	UV (230 nm)	1.0	10	2.05	35
Dimethomorph	UV (230 nm)	1.0	10	3.67	35
Myclobutanil	UV (230 nm)	1.0	10	4.62	35
Difenoconazole	UV (230 nm)	1.0	10	8.65	35
Nitrates	El. Conductivity	1.0	1000	13.0	30
Phosphates	El. Conductivity	1.0	1000	16.8	30
Potassium	El. Conductivity	1.0	1000	5.6	30

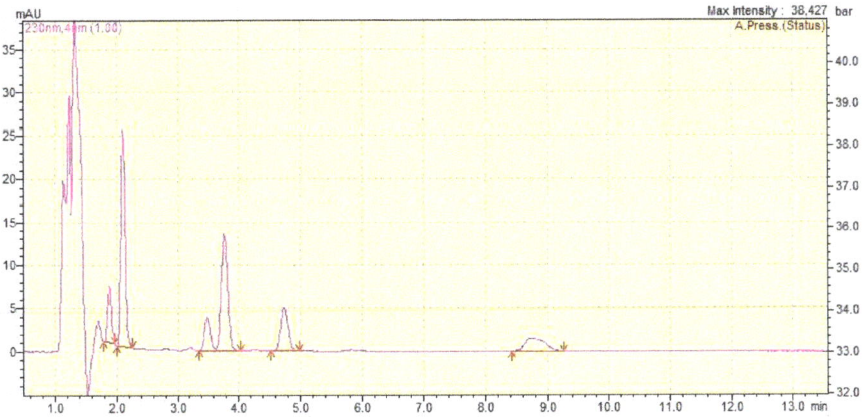

Figure 2. HPLC chromatogram representing the analyte peaks (as reported in Table 2).

2.4. Estimation of Pollutant Reductions

The reductions of nutrients and herbicides were calculated considering the measured concentrations in both the influent (C_{inlet}) and the respective effluent (C_{outlet}) for each tank, for each sampling timepoint, according to the following equation:

$$\%\text{Reduction} = \frac{C_{inlet} - C_{outlet}}{C_{inlet}} 100 \quad (1)$$

Mean reductions for every pollutant and system were also calculated using Equation (1) and considering the average inlet and the average outlet concentrations for the total period of monitoring.

2.5. Statistical Analysis

A statistical analysis was performed to investigate the significance of the derived results. The tests that were undertaken were: (a) a single-factor analysis of variance (ANOVA), in order to investigate whether the derived differences in the removal rates among the four experimental tanks are significant; (b) a Tukey's Honest Significant Difference (HSD) post hoc test so as to identify between which particular groups (examined tanks) these differences were significant; and (c) a Student's *t*-test in order to examine the discrepancies in pollutant removal rates per aquatic plant system, between the summer and winter period. All tests were applied in MS Excel 2016, using the data analysis toolpack and considering a significance level of 0.05. Moreover, we calculated the Pearson correlation coefficients between each recorded meteorological variable and the removal rates derived for each of the planted and the reference tank, respectively. The examined meteorological variables were the: air temperature, solar radiation, wind speed and precipitation. The above were exploited to derive estimates of potential evaporation (for the reference tank) and evapotranspiration (for the planted tanks) using the Penman (1948) and Penman–Monteith (Monteith, 1965) methods [75,76], respectively.

3. Results and Discussion

3.1. Physicochemical Parameters

The statistics of pH, Electrical Conductivity, Total Dissolved Solids, Salinity and water temperature (i.e., mean, standard deviation, minimum and maximum values) in the four different tanks, calculated for the entire operation period, are presented in Table 3. The temporal variation of the physicochemical parameters in the three experimental tanks, as well as at their inlet, measured during loading of the wastewater, is illustrated in Figures 3 and 4.

Table 3. Physicochemical parameters in the test system tanks.

Statistical Parameter	T (°C)				Salinity (ppt)				pH			
	Hyac.	Azola	Lemna	Contr.	Hyac.	Azola	Lemna	Contr.	Hyac.	Azola	Lemna	Cont.
Minimum	19.0	19.1	19.0	19.0	0.2	0.2	0.4	0.4	6.8	6.9	7.1	7.3
Mean	26.3	26.8	26.9	27.1	0.6	0.6	0.6	0.5	7.7	8.0	8.1	8.5
Maximum	32.2	32.4	32.5	32.2	0.8	0.8	0.7	0.6	8.5	8.6	8.7	9.0
St. Dev.	4.01	3.81	3.91	3.66	0.14	0.09	0.07	0.06	0.40	0.39	0.37	0.38
	EC (μS/cm)				TDS (mg/L)							
	Hyac.	Azola	Lemna	Contr.	Hyac.	Azola	Lemna	Contr.				
Minimum	394.7	306.1	771.0	737.0	233.4	221.0	559.0	513.5				
Mean	1300.3	1301.5	1181.4	1044.9	840.0	818.3	740.3	648.0				
Maximum	1845.0	1560.0	1479.0	1297.0	1087.5	992.0	886.5	825.5				
St. Dev.	306.30	209.43	157.19	130.48	176.85	114.28	87.81	77.57				

3.2. Nutrients

Influent and effluent concentration and removal statistics for the overall experimental period and for all analytes were estimated and are presented in Table 4. The mean reductions were calculated considering the average influent concentration and the respective effluent levels per pollutant.

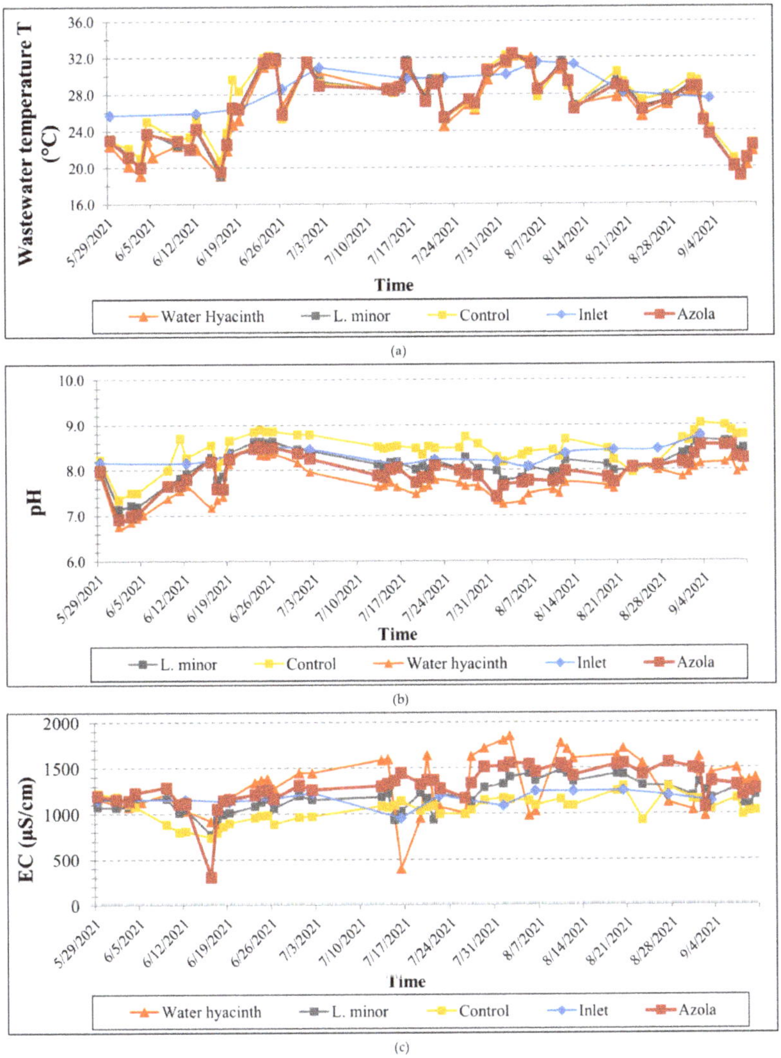

Figure 3. Temporal variation of (**a**) wastewater temperature, (**b**) pH and (**c**) electrical conductivity at the inlet and outlet of the four tanks throughout the operation period.

Table 4. Removal percentages for the examined systems.

Pollutant	System			
	Azola	Water Hyacinth	Lemna	Control
		NH_4^+-N (%)		
Mean	84.3	91.9	84.7	41.8
Max	100.0	100.0	100.0	79.6
Min	24.8	38.3	10.9	0.0
		PO_4^{3-}-P (%)		
Mean	68.5	65.0	77.2	76.2
Max	78.8	81.9	91.6	98.1
Min	25.3	20.6	45.5	36.4

Table 4. *Cont.*

Pollutant	System			
	Azola	Water Hyacinth	Lemna	Control
		NO_3^--N (%)		
Mean	6.6	20.4	76.1	35.8
Max	94.4	50.0	98.8	98.9
Min	0.0	0.0	0.0	0.0
		Imidacloprid (%)		
Mean	31.1	34.7	43.3	68.3
Max	75.0	81.7	81.9	88.3
Min	0.0	0.0	0.0	0.0
		Thiacloprid (%)		
Mean	6.6	34.5	12.6	0.9
Max	81.9	83.3	81.9	63.8
Min	0.0	0.0	0.0	0.0
		Dimethomorph (%)		
Mean	22.4	41.7	32.8	53.1
Max	89.2	82.9	80.0	75.7
Min	0.0	0.0	0.0	0.0
		Myclobutanil (%)		
Mean	6.0	19.5	7.5	34.9
Max	76.7	79.7	77.1	56.3
Min	0.0	0.0	0.0	0.0
		Difenoconazole (%)		
Mean	33.2	64.8	33.8	69.0
Max	92.0	94.6	82.9	95.6
Min	0.0	20.40	0.0	20.4

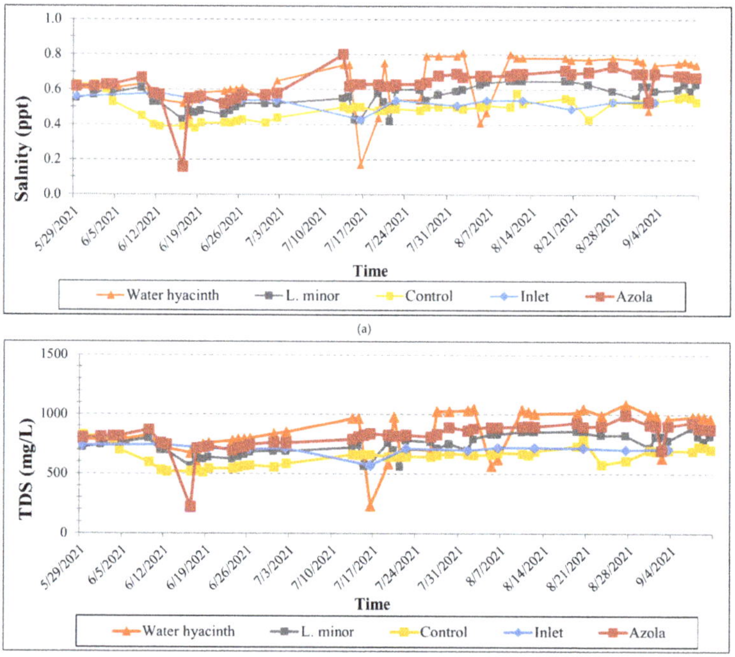

(a) (b)

Figure 4. Temporal variation of (a) salinity and (b) total dissolved solids at the inlet and the outlet of the four tanks throughout the operation period.

The temporal variation of ammonium, nitrate and phosphate concentrations is presented in Figure 5. Accordingly, the temporal variations of the weekly removal rates per aquatic plant and pollutant, in parallel with the wastewater temperature, are shown in Figure 6.

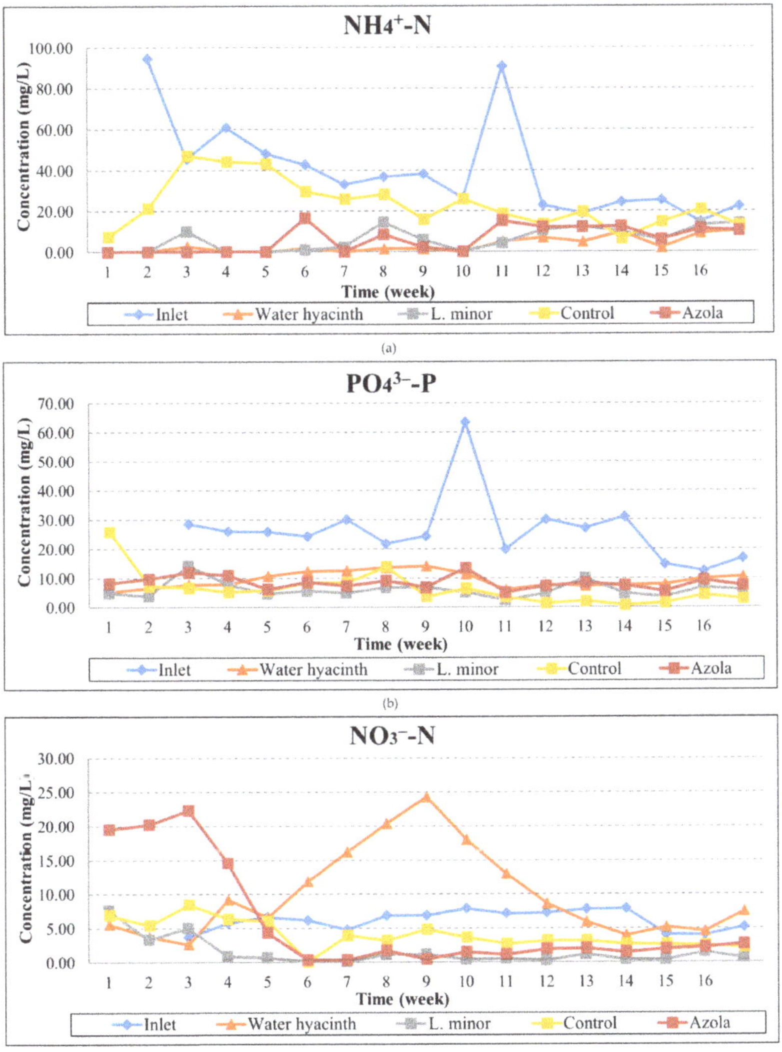

Figure 5. Temporal variation of: (**a**) NH_4^+-N, (**b**) PO_4^{3-}-P, (**c**) NO_3^--N concentrations at the inlet and the outlet of the three tanks throughout the operation period.

As can be seen from the graphs, ammonium ions exhibited reduction percentages ranging from 38.3% to 100% for water hyacinth, 11% to 100% for *L. minor*, 24.8% to 100% for *A. pinnata*, and up to 79% for the control tank. It is, therefore, noticed that the water hyacinth presented the highest reduction between the three aquatic macrophytes. The mean removal rates for the whole experiment period were thus 91.9% for water hyacinth, 84.3% for azola, 84.7% for lemna and 41.8% for the control tank.

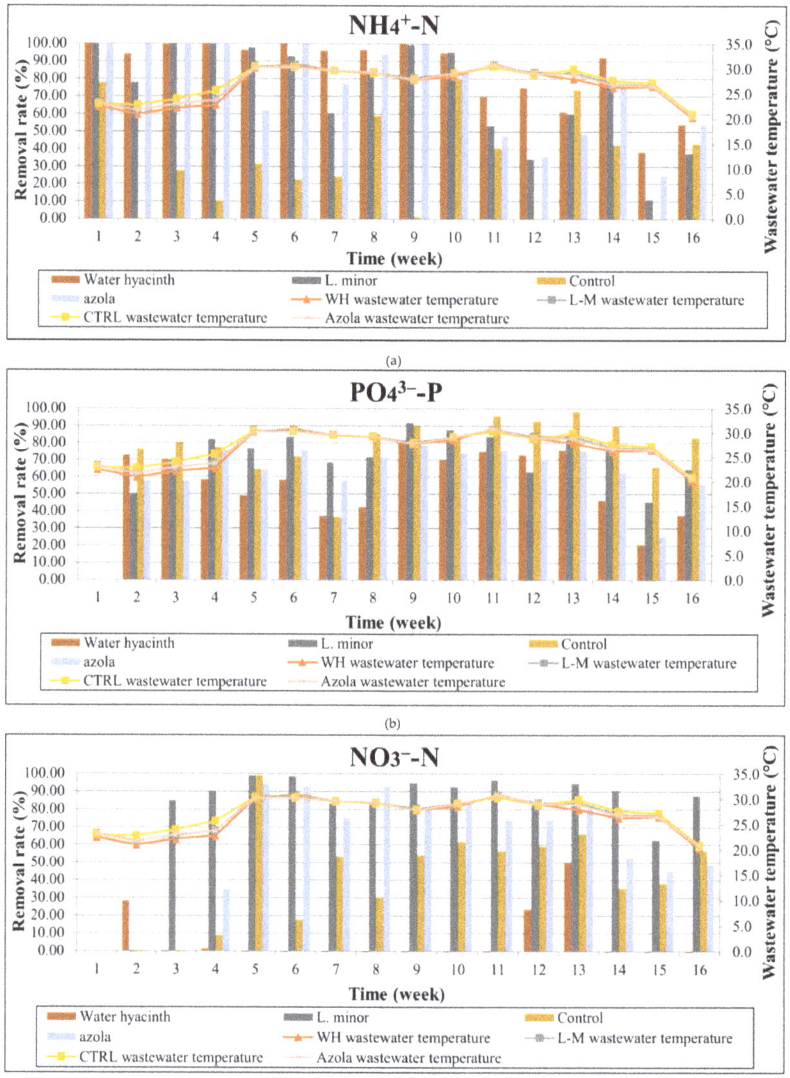

Figure 6. Temporal evolution of removal rate per examined tank and wastewater temperature for: (**a**) NH_4^+-N, (**b**) PO_4^{3-}-P, and (**c**) NO_3^--N.

Phosphate ions also presented significant reductions that ranged from 20.6% to 81.9% for water hyacinth, 45.5% to 91.6% for lemna, 25.3% to 78.8% for azola and between 36.6% to 98% for the algae control system, values demonstrating that the presence of algae is important for the phosphates' uptake, as well as the fact that phosphorus is absorbed in the aquatic plants' roots and surfaces and potentially re-dissolves during sampling and feeding procedures. The mean reductions during the 4-month experimental period ranged between 65% and 77.2%, with the minimum observed for hyacinth and the maximum for *Lemna minor*.

Finally, NO_3^--N presented reductions ranging up to 50% for the water hyacinth system, up to 98.8% for the lemna system, up to 94.4% for the azola system and up to 98.9%

reduction for the control tank. As a general observation, the lemna and azola systems presented a higher consistency in nitrate removal, contrary to the water hyacinth system where high variations were observed between samplings. An increase in the detected nitrate concentrations, along with an accumulation effect, was apparent from the 6th and 11th weeks of the experiment. As regards the mean reductions, they ranged from no reduction and up to 76.1%, the latter observed for the lemna system.

Our findings are generally supported by previous ones, or even presented slightly better pollution reduction potential. Comparing the present experiment performance with our previous relevant study (performed during winter; [2]) it can be concluded that reductions were comparable. In more detail, phosphate reduction was between 61.2–99.6% for lemna and 64.4–98.2% for water hyacinth, and nitrate reduction was between 18–78.4% for lemna and 19.5–78.4% for water hyacinth [2]. Using the same rationale, Sarkheil and Safari [37] observed reductions of ammonium and phosphate ions by 43.7% and 52.4%, respectively, using *Lemna minor*. In an identical combination to ours, that is, of *Lemna minor* and algae, reductions of 12.5% and 70.47% were observed for nitrates and phosphates, respectively [40], whereas Sudiarto et al. [41] presented 36.2% phosphorus uptake by lemna from livestock wastewater.

Accordingly, Fox et al. [53] reported nitrogen removals up to 85%, after a 4-week treatment using water hyacinth for N-concentrations up to 300 ppm. Similarly, Nabi et al. [54] reported water hyacinth capability for removal of 63.3% and 58.5% for TN and P, respectively, while Osti et al. [55] reported a reduction in TN, TIN, TP and PO_4^{3-}-P, equal to 66%, 82%, 27% and 33%, respectively. Significant reductions reaching 87% for nitrogen compounds and 85% for total phosphorus were also observed by Ozengin and Elmaci [77]. Wang et al. [78] reported 63.3% reduction of phosphates from a contaminated river system using water hyacinth, whilst Qin et al. [79] presented TN and TP reductions of 47.4% and 53.4%, respectively. Accordingly, Kumari and Tripathi [80] presented elimination of 26.6% of NO_3^--N, 53.0% of Total Kjeldahl Nitrogen (TKN) and 56.6% of PO_4^{3-}-P concentrations.

At a different spatial scale, Wang et al. [61] studied nitrogen pollution reduction in a Chinese lake, with the results presenting reduction of 52–64% using water hyacinth. High N reductions (66–82%) and relatively lower ones for P (27–33%) were observed by Osti et al. [55] in fishponds with water hyacinth, compared to no-plant ponds, whereas in a sewage treatment study, Aremu et al. [59] reported that after a 28-day experimental period, the water hyacinth cultured sewage had reduced 45.5% of nitrate and 37.8% of phosphorus, whereas the pH also dropped from 8.6 to 7.8. Moreover, reductions of 67% of phosphorus and 92% of nitrate using water hyacinth as a treatment plant were reported by Kutty et al. [56]. Remarkable removal efficiencies of TN, P and K from wastewater that reached 63.28%, 58.54% and 85.89%, respectively, were noticed by Nabi et al. [54]. Xu et al. [81] reported removal rates of NH_4^+-N, NO_3^--N, NO_2^--N, TN, TP, COD, and Chlorophyl-a ranging between 26.4% (ultimate minimum, only for nitrites) and 99.5%. Finally, slightly lower reductions were presented by Zhao et al. in a relevant study design, where a minimum reduction of 25.6% and a maximum of 64.5% were mentioned [60].

3.3. Pesticides

Influent and effluent concentrations and removal statistics for the overall experimental period and for all analytes were estimated and are presented in Table 4.

The temporal variation and removal rates are presented in Figures 7 and 8, respectively. Imidacloprid presented removals ranging from no reduction to 81.7% for water hyacinth, no removal to 81.9% for Lemna minor, no removal to 75% for Azolla pinnata and no removal to 88.3% for the algae control. Thiacloprid showed slightly better reductions in all aquatic plant systems, reaching up to 83.3% removal for the hyacinth system, 81.9% for the lemna and azola systems, and 63.8% for the algae control tank. Accordingly, dimethomorph presented reductions from no removal to 82.9% in the water hyacinth system, no removal to 80% for the lemna system, no removal to 89.2% in the azolla system and up to 75.7% for the algae control. Finally, difenoconazole exhibited disappearances up to 94.6% for the water hyacinth, 82.9% for lemna, 92% for azola and 95.6% for the algae control tank, thus proving that for this compound, the aquatic plants presence did not play a significant role, possibly due to their intrinsic environmental (DT50) or physicochemical properties (hydrolysis, photolysis).

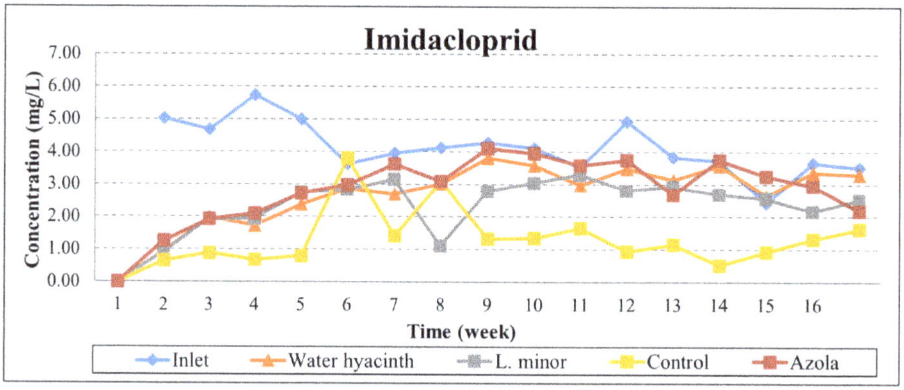

(a)

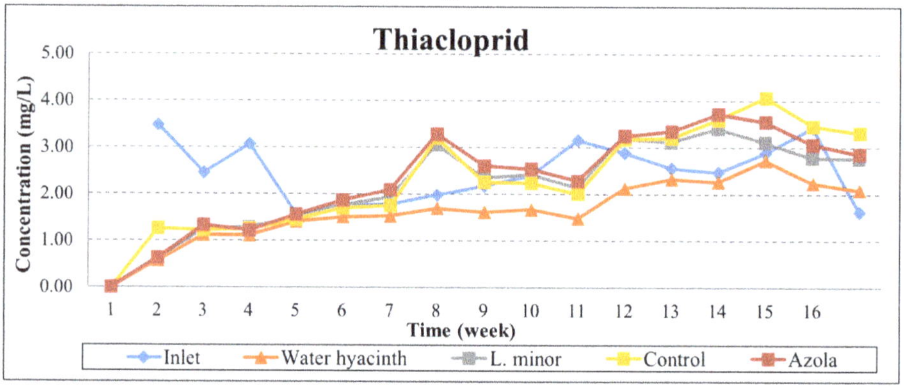

(b)

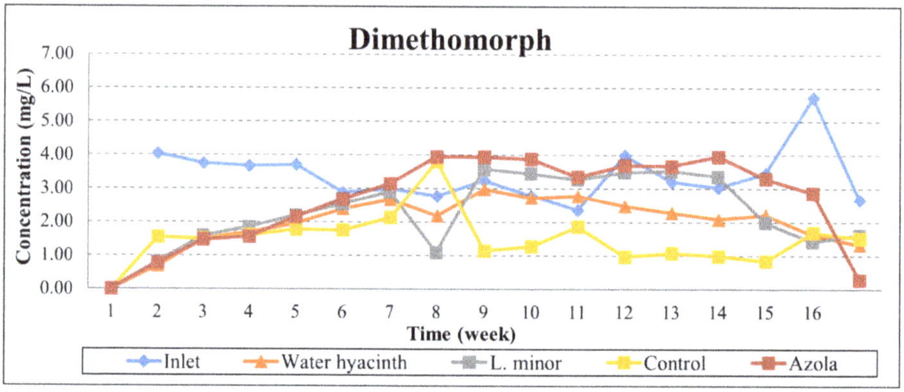

(c)

Figure 7. *Cont.*

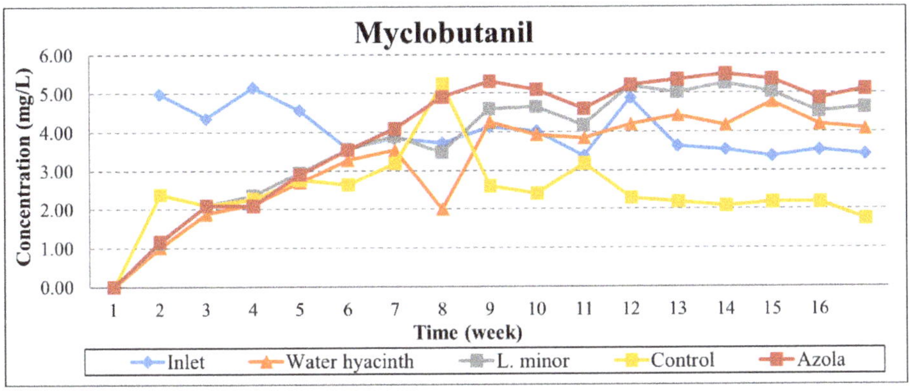

(d)

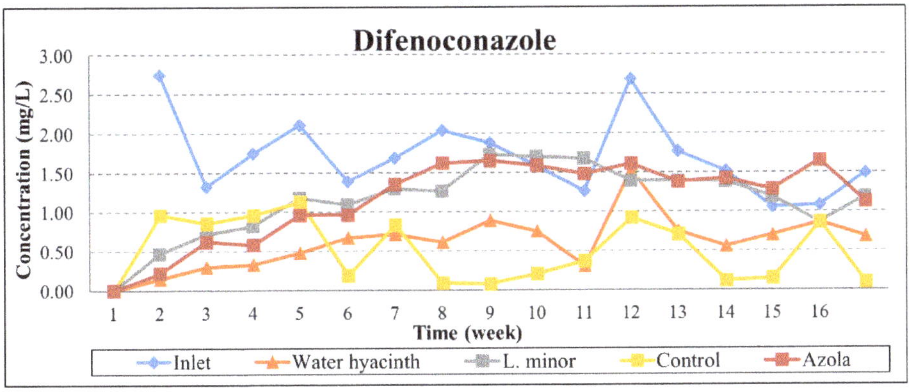

(e)

Figure 7. Temporal variation of: (**a**) Imidacloprid, (**b**) Thiacloprid, (**c**) Dimethomorph, (**d**) Myclobutanil and (**e**) Difenoconazole concentrations at the inlet and outlet of the three tanks throughout the operation period.

The mean imidacloprid reductions for the whole study period were 31.1% for azola, 34.7% for water hyacinth and 43.3% for the lemna system. At the same rationale, thiacloprid was reduced by 6.6%, 34.5% and 12.6% for the respective systems, and dimethomorph reductions were 22.4%, 41.7% and 32.8% for azola, hyacinth, and lemna, respectively. Finally, as regards myclobutanil, the mean reductions were 6%, 19.5% and 7.5% for azola, hyacinth and lemna, respectively, whilst difenoconazole was respectively reduced by 33.2%, 64.8% and 33.8% throughout the experimental period.

Pesticide reductions with Lemna minor were also examined by Dosnon–Olette et al. [42], where a 17% reduction was demonstrated for dimethomorph. Similarly, dimethomorph exhibited up to 60% reduction in the study of Ekperusi et al. [82] with Lemna minor, yet with higher pesticide concentrations, reaching 1 mg/L; however, no control system was run in parallel. Additionally, in our previous experiment, under winter incubation, the reductions achieved by Lemna minor were from 10.4% to 49.9% for imidacloprid, no reduction and up to 38.8% for thiacloprid, 13.2% to 63.5% for dimethomorph and 0.8% to 60.8% for myclobutanil, with the maximum obtained for temperatures above 15 °C [2].

Regarding the degradation using water hyacinth, as reported in our previous study, the removal rates for water hyacinth treatment were from 11.4% to 65.6% for imidacloprid, no reduction and up to 57.8% for thiacloprid, 3.6% to 74.1% for dimethomorph and 4.2% to 65.1% for myclobutanil, with the maximum obtained also in this case for temperatures above 15 °C [2]. To our knowledge, there is currently no other study available in the literature considering this specific plants–pesticides combination; hence, no further comparison is feasible.

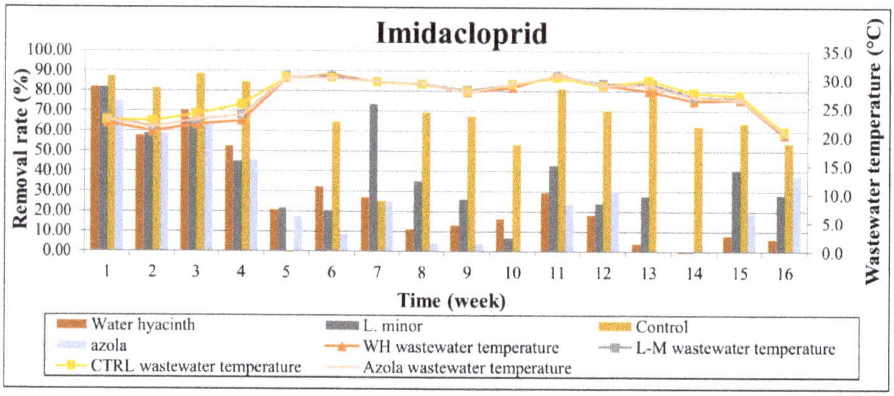

(a)

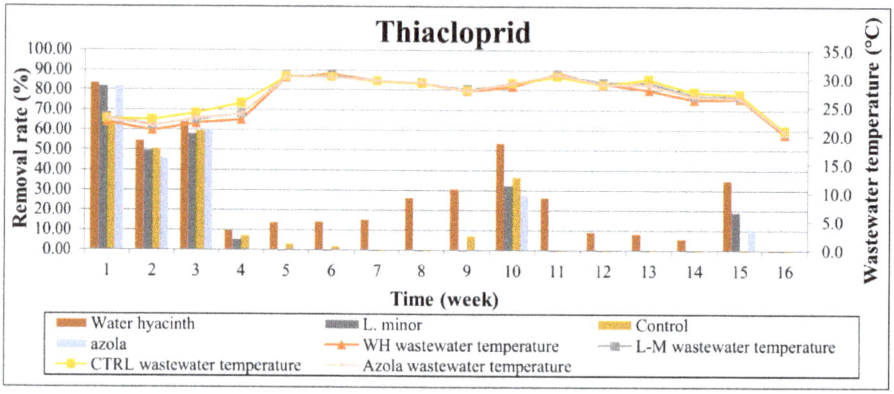

(b)

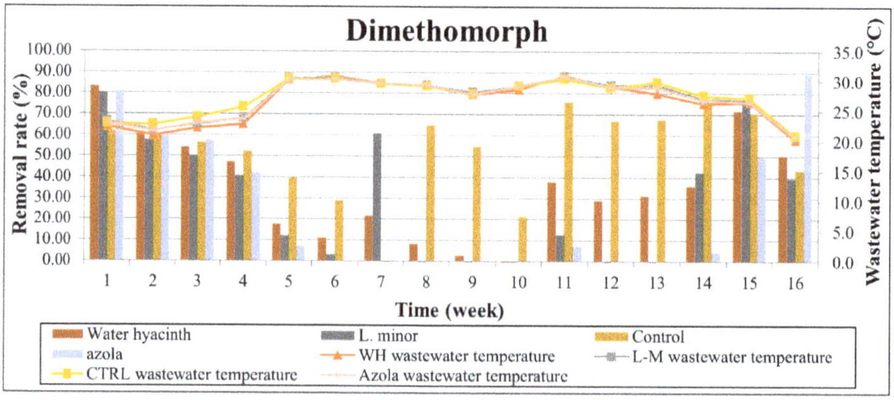

(c)

Figure 8. *Cont.*

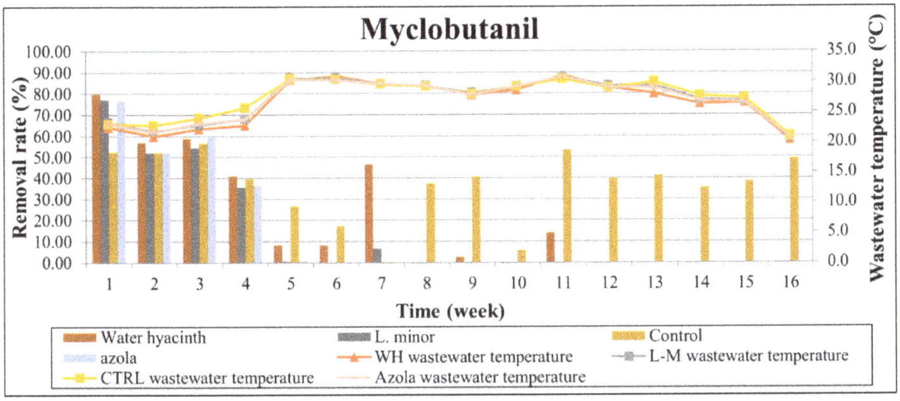

(d)

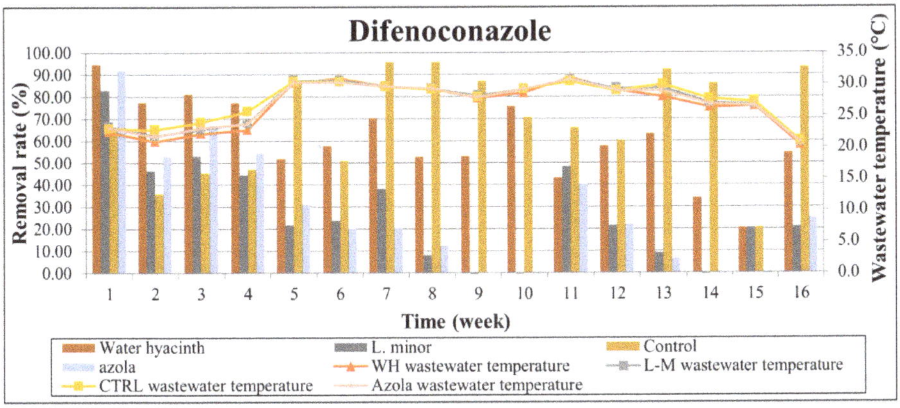

(e)

Figure 8. Temporal variation of removal rate per examined tank and wastewater temperature for: (**a**) Imidacloprid, (**b**) Thiacloprid, (**c**) Dimethomorph, (**d**) Myclobutanil, and (**e**) Difenoconazole.

Algae that developed in the control seem to have had a positive effect in reducing pesticides, a fact that has been previously recognized in the literature [83–86]. Riaz et al. [87] also reported almost equal reductions using water hyacinth and algae for pesticide phytoremediation. Nevertheless, algae cannot be easily adopted in CFW systems as their growth is, in most cases, uncontrollable, causing system clogging and malfunctions, and at the same time creating odor and appearance issues in contrast to aquatic plant installations. In parallel, other limitations of algae systems include pH limitations, zooplankton and herbivorous protozoa contamination, need of wastewater pretreatment, and the need to establish mixed cultures to achieve maximum removal efficiency which requires special attention in order to avoid inter-species competition [84].

3.4. Meteorological Conditions

The most important meteorological parameters associated with plant growth and agrochemical decomposition, that is, air temperature, solar radiation, precipitation and wind speed, were recorded at the experimental site and are presented in Figure 9. Besides, evaporation and reference evapotranspiration rates, for the control and planted tanks, respectively, were also estimated at a daily time step, making use of the meteorological parameter records acquired from the meteorological station. The daily estimations were then averaged to give average weekly rates. The calculations were performed based on

Penman and Penman–Monteith methodologies [75,76], for evaporation from open water (E_0) and potential evapotranspiration from a reference crop (ET_0), respectively. E_0 and ET_0 variation throughout the operation period are depicted in Figure 9c.

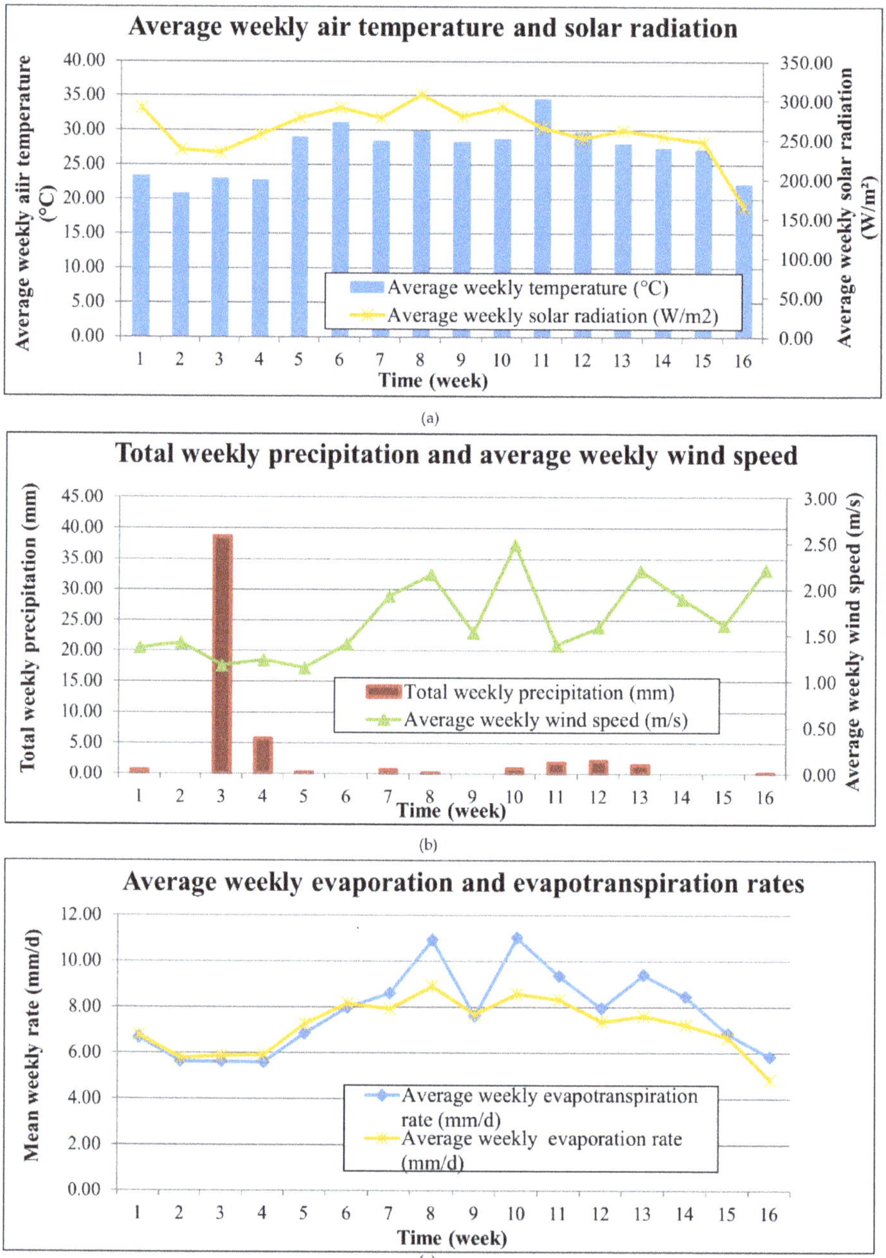

Figure 9. Temporal variation of (**a**) average weekly air temperature and solar radiation; (**b**) total weekly precipitation and average weekly wind speed; and (**c**) average weekly evaporation and evapotranspiration rates.

The minimum weekly average air temperature was found to be 20.8 °C and the maximum 34.5 °C. The mean daily temperature for the study period was 27.1 °C, whereas the overall minimum and maximum temperatures recorded were 14.3 °C and 42.7 °C, respectively.

3.5. Statistical Analysis Results

As mentioned, an ANOVA test followed by a Tukey's Honest Significant Difference (HSD) post hoc test were implemented to investigate the statistical significance of the experimental data. To perform the ANOVA test, four groups consisting of the removal rates for each of the four experimental tanks were considered. The analysis revealed significant differences ($p \leq 0.05$) in the performance of the examined systems for almost all examined pollutants except thiacloprid. Therefore, for the rest of nutrients and herbicides, the Tukey Honest Significant Difference (HSD) post hoc test was also applied by comparing, for each pollutant, all possible pairs of groups (experimental tanks). With respect to the ammonium ions, Tukey's test demonstrated a significantly higher performance of all aquatic plant systems compared to the algae control tank. Although water hyacinth presented relatively higher reduction rates compared to the other species, no statistically significant difference was detected among the planted tanks. Regarding phosphates, the results showed a significantly lower performance of the water hyacinth system compared to the *Lemna minor* and the algae control ones. As regards nitrates, significant differences in the reduction rates emerged between water hyacinth and *Lemna minor* and water hyacinth and control tank with the water hyacinth exhibiting a lower performance in both cases. As far as pesticides removal is concerned, the algae control tank presented a significantly better performance compared to all planted systems with respect to imidacloprid reduction, whereas in the case of dimethomorph, the latter yielded significantly higher removal rates only compared to azola. A significant difference was also identified with respect to myclobutanil removal rates between control tank and azola and *Lemna minor* aquatic systems, with the former presenting significantly higher efficiency in both cases. Finally, for the last compound, significant differences in the difenoconazole removal rates emerged between: (a) water hyacinth (higher) and azola; (b) control tank (higher) and azola; (c) water hyacinth (higher) and *L. minor*; and (d) control tank (higher) and *Lemna minor*. Table 5 presents the ANOVA results per pollutant.

Table 5. Results of ANOVA test performed between the four experimental tanks, that is, azola, water hyacinth, *Lemna minor* and control tank, for each of the examined pollutants.

Pollutant	F	p-Value	F Crit
NH_4^+-N *	12.990	1.21×10^{-6}	2.758
PO_4^{3-}-P *	5.555	0.002	2.769
NO_3^--N *	6.237	0.001	2.769
Imidacloprid *	8.370	9.82×10^{-5}	2.758
Thiacloprid	2.196	0.098	2.758
Dimethomorph *	3.322	0.026	2.758
Myclobutanil *	4.527	0.006	2.758
Difenoconazole *	12.376	2.10×10^{-6}	2.758

* Statistically significant difference ($p \leq 0.05$).

Correlation coefficients between pollutant reduction rates and meteorological variables were also determined in an effort to detect eventual dependencies between the results and meteorological conditions. The Pearson coefficient values are shown in Table 6, with those revealing a strong linear relationship ($r > 0.5$), either positive or negative, being depicted in bold.

As a general observation, the meteorological variables that affected the efficiency of the examined systems the most are the air temperature, the evapotranspiration from the planted tanks and the wind speed, whereas precipitation did not present any significant correlation with the removal rate results, probably due to the negligible rainfall depth that was recorded during the summer period. Based on the derived correlation values, it could be concluded that the examined herbicides seem to present a more consistent behavior, that is, a negative correlation with the examined meteorological variables for most of the compounds and examined tanks. Some exceptions that are observed (positive correlation values), mainly for the cases of thiacloprid and myclobutanil, are rather low, and thus, do not indicate strong dependencies.

Table 6. Pearson correlation coefficients between meteorological variables and pollutant reduction rates per pollutant and examined tank. Bold values indicate a strong linear relationship.

Examined Tank	Pollutant							
	NH_4^+-N	PO_4^{3-}-P	NO_3^--N	Imidacroprid	Thiacloprid	Dimethomorph	Myclobutanil	Difenoconazole
	Pearson Correlation Coefficients between Temperature and Pollutant Reduction							
Azola	−0.352	0.347	**0.648**	**−0.607**	−0.357	**−0.718**	**−0.511**	−0.402
Water hyacinth	−0.082	0.096	−0.364	−0.480	−0.197	**−0.591**	−0.450	−0.497
Lemna	−0.149	0.441	**0.538**	−0.408	−0.342	**−0.575**	**−0.515**	−0.356
Control	0.098	0.129	**0.594**	−0.251	−0.202	−0.044	−0.308	0.276
	Pearson correlation coefficients between solar radiation and pollutant reduction							
Azola	0.340	0.441	0.329	−0.289	0.250	**−0.622**	0.081	−0.049
Water hyacinth	**0.539**	0.169	**−0.545**	0.073	0.420	−0.478	0.135	0.122
Lemna	0.486	0.471	0.243	−0.013	0.289	−0.362	0.127	−0.129
Control	0.266	−0.111	0.196	−0.103	0.431	−0.163	−0.373	0.133
	Pearson correlation coefficients between evapotranspiration/evaporation and pollutant reduction							
Azola	−0.128	0.402	**0.505**	**−0.671**	−0.281	**−0.782**	**−0.560**	−0.457
Water hyacinth	0.022	0.091	−0.368	**−0.514**	−0.038	**−0.690**	−0.456	−0.210
Lemna	−0.013	0.439	0.368	−0.411	−0.247	**−0.652**	**−0.514**	**−0.577**
Control	0.317	0.053	0.413	−0.179	0.041	−0.170	−0.445	0.350
	Pearson correlation coefficients between wind speed and pollutant reduction							
Azola	−0.212	0.050	0.299	**−0.523**	−0.476	−0.392	**−0.692**	**−0.537**
Water hyacinth	−0.297	−0.184	−0.114	**−0.624**	−0.299	−0.422	**−0.559**	−0.131
Lemna	−0.278	0.097	0.143	−0.398	−0.454	−0.385	**−0.619**	**−0.650**
Control	0.473	0.142	0.284	−0.078	−0.496	−0.206	−0.331	**0.532**
	Pearson correlation coefficients between precipitation and pollutant reduction							
Azola	0.238	−0.111	−0.275	0.423	0.368	0.301	0.438	0.381
Water hyacinth	0.174	0.203	0.096	0.466	0.308	0.225	0.369	0.333
Lemna	0.236	−0.047	0.071	0.335	0.334	0.191	0.398	0.305
Control	−0.020	0.039	−0.213	0.293	0.336	0.067	0.271	−0.272

On the other hand, with respect to nutrient removal, it can be seen that there is a diverse image in the way the efficiency of the systems is affected by the examined meteorological variables, with correlation values sometimes revealing a positive relationship and sometimes a negative one, depending on the considered tank and pollutant. These results are in contrast with our previous relevant study [2], performed during winter, where a consistently positive correlation had been revealed between the examined system efficiency and temperature, radiation and evapotranspiration. This disagreement may indicate, especially for temperature, that there is a certain upper temperature threshold above which the aquatic plant system performance is reduced. The main reason for this is probably the increase in the concentrations of the various substances due to increased evapotranspiration and resulting water loss.

Finally, an attempt was made to compare the pollutant removal performance of the examined systems during summer (current experiment) and winter period, when a similar experiment had been elaborated [2]. For this purpose, we considered a 2-month period, from November 2020 until early January 2021 where the average wastewater temperatures were below 15 °C using removal rates by Pavlidis et al. [2], and compared with respective data from the current experiment, undertaken during the summer season, when the wastewater in the experimental tanks had an average temperature of approximately 27 °C. To analyze the differences in the results between the two experiments, we performed a two-sample Student's t-test by examining each combination of experimental tank and pollutant separately. It should be noted that only the compounds and experimental systems that had been considered in both experiments were included in the analysis, namely, water hyacinth, *L. minor* and the control tank with respect to the examined systems, and phosphates, nitrates, imidacloprid, thiacloprid, dimethomorph and myclobutanil with respect to the considered pollutants. The results revealed a significant difference ($p = 0.001$) between the winter and summer periods for the case of water hyacinth and phosphate and nitrate removals, indicating a significantly higher level of performance of the specific aquatic plant for both compounds during the winter. On the contrary, a significantly higher ($p = 0.00008$) removal performance with respect to nitrates was seen for *L. minor* during the summer period. Regarding pesticide reduction, the examined aquatic plant systems did not show any significant difference in their performance between summer and winter. Some significant discrepancies were only derived for the control tank and for three of the examined pesticides, that is, imidacloprid ($p = 0.00002$), dimethomorph ($p = 0.0003$), and myclobutanil ($p = 0.01$). For these three compounds, the analysis demonstrated a significantly higher efficiency of the algae control tank during the current experiment (summer season).

4. Conclusions

The necessity, significance, and efficiency of natural treatment and remediation systems has been previously reported in various studies dealing with domestic wastewater treatment, with regards to organic matter, nitrogen and phosphorus removal. The constructed floating wetlands (CFW) that were also examined in the present study constitute an efficient, easily applicable and low-cost treatment option. In this context, the efficiency of three different constructed floating wetlands pilot-scale systems consisting of aquatic macrophytes as a pollution control plant, in the removal of nutrients and five pesticides was examined in comparison with an unplanted system over a 16-week period during spring to summer of 2021, indicating very promising results. Reductions reaching almost 100% were found for all examined agrochemicals. The highest mean reduction percentages were observed for phosphates, ammonium, nitrates and difenoconazole.

Both for ammonium, nitrates and thiacloprid, the aquatic macrophyte systems performed better than the control. On the contrary, potassium presented high residues, possibly due to adsorption and accumulation to plant roots, with subsequent re-suspension in the water column. Moreover, for the rest of the pesticides, the reductions observed in the plant systems were comparable to those of the control system, demonstrating that the effect of aquatic plants was comparable to the one provided by the green algae which developed in the control tank. It shall also be remarked that the potential for plant uptake and absorption to aquatic plant surfaces cannot be neglected based on the pesticide physicochemical properties. An exception was observed for imidacloprid, where photolytic degradation was the prevailing dissipation process, which was rather expectable based on the photolytic half-life of the compound (1 h; Table 1).

Overall, the statistical analysis results revealed a greater dependency of the pesticides removal potential on the examined meteorological parameters, especially temperature, wind speed and evapotranspiration. In particular, an increase in these variables proved to negatively affect the aquatic plant system efficiency (negative correlation), which is in contradiction with our previously published results derived from a similar experiment running during the autumn to winter period. This may be explained by the subsequent reduction in water volume in the tanks due to the high evapotranspiration rates, and therefore, the consequent increase in pollutant concentrations. Besides, the differentiated results between the two periods demonstrate the existence of specific threshold values of the examined variables, especially air temperature, above which pesticide decontamination starts being hampered. Additional experimentation under longer and/or different periods (e.g., spring season) characterized by different temperature ranges is required to derive generalized results.

It can be concluded that the examined systems under the Mediterranean geoclimatic conditions described in the present study have the potential for decontamination of spray tank mix remnants, or pesticide container wash-off water, as well as agricultural runoff from agricultural field drainage networks, thus consisting of a valuable technology for farmers and policy-makers. Further research is deemed necessary, considering different pollutant and plant combinations, variation of HRT, analysis of aquatic macrophytes to establish mass balance, as well as the correlation of pollution reduction with chlorophyll content and other plant-related parameters.

Author Contributions: Conceptualization, G.P. and V.A.T.; Data curation, G.P., I.Z., H.K. and A.M.; Formal analysis, G.P., I.Z., H.K. and V.A.T.; Funding acquisition, G.P. and V.A.T.; Investigation, G.P., I.Z., G.B. and A.M.; Methodology, G.P., I.Z. and V.A.T.; Project administration, V.A.T.; Resources, H.K. and V.A.T.; Supervision, V.A.T.; Validation, I.N., H.K. and V.A.T.; Writing—original draft, G.P. and I.Z.; Writing—Review & Editing, H.K., I.N. and V.A.T. All authors have read and agreed to the published version of the manuscript.

Funding: The study was carried out and funded in the context of the project "Investigation of the efficiency of natural treatment systems in the removal of pesticides and fertilizers" (MIS 5049184) under the call for proposals on "Researchers' support with an emphasis on young researchers—second Cycle". The project was co-financed by Greece and the European Union (European Social Fund-ESF) through the Operational Programme Human Resources Development, Education and Lifelong Learning 2014–2020.

Institutional Review Board Statement: Not applicable.

Informed Consent Statement: Not applicable.

Data Availability Statement: All data are included in the paper.

Conflicts of Interest: The authors declare no conflict of interest.

References

1. Zotou, I.; Tsihrintzis, V.A.; Gikas, G.D. Performance of seven water quality indices (WQIs) in a Mediterranean River. *Environ. Monit. Assess.* **2019**, *191*, 505. [CrossRef] [PubMed]
2. Pavlidis, G.; Zotou, I.; Karasali, H.; Marousopoulou, A.; Bariamis, G.; Tsihrintzis, V.A.; Nalbantis, I. Performance of pilot-scale constructed floating wetlands in the removal of nutrients and pesticides. *Water Resour. Manag.* **2022**, *36*, 399–416. [CrossRef]
3. Pavlidis, G.; Tsihrintzis, V.A. Environmental benefits and control of pollution to surface water and groundwater by agroforestry systems: A review. *Water Resour. Manag.* **2018**, *32*, 1–29. [CrossRef]
4. Ladislas, S.A.; El-Mufleh, A.C.; Gerente, C.F.; Chazarenc, F.Y.; Andres, Y.B.; Bechet, B. Potential of aquatic macrophytes as bioindicators of heavy metal pollution in urban stormwater runoff. *Water Air Soil Pollut.* **2012**, *223*, 877–888. [CrossRef]
5. Stefanakis, A.; Akratos, C.S.; Tsihrintzis, V.A. *Vertical Flow Constructed Wetlands: Eco-Engineering Systems for Wastewater and Sludge Treatment*; Elsevier: Amsterdam, The Netherlands, 2014; 378p.
6. Alemu, T.; Lemma, E.; Mekonnen, A.; Leta, S. Performance of pilot scale anaerobic-SBR system integrated with constructed wetlands for the treatment of tannery wastewater. *Environ. Process.* **2016**, *3*, 815–827. [CrossRef]
7. Pavlineri, N.; Skoulikidis, N.T.; Tsihrintzis, V.A. Constructed floating wetlands: A review of research, design, operation and management aspects, and data meta-analysis. *Chem. Eng. J.* **2017**, *308*, 1120–1132. [CrossRef]
8. Mustafa, H.M.; Hayder, G. Recent studies on applications of aquatic weed plants in phytoremediation of wastewater: A review article. *Ain Shams Eng. J.* **2020**, *12*, 355–365. [CrossRef]
9. Sharma, S.; Singh, B.; Manchanda, V.K. Phytoremediation: Role of terrestrial plants and aquatic macrophytes in the remediation of radionuclides and heavy metal contaminated soil and water. *Environ. Sci. Pollut. Res.* **2015**, *22*, 946–962. [CrossRef]
10. Angassa, K.; Leta, S.; Mulat, W.; Kloos, H.; Meers, E. Evaluation of pilot-scale constructed wetlands with *Phragmites karka* for phytoremediation of municipal wastewater and biomass production in Ethiopia. *Environ. Process.* **2019**, *6*, 65–84. [CrossRef]
11. Pavlidis, G.; Karasali, H. Natural Remediation Techniques for Water Quality Protection and Restoration. In *Methods for Bioremediation of Water and Wastewater Pollution*; Springer: Cham, Switzerland, 2020; pp. 327–340.
12. Kadlec, R.H.; Wallace, S. *Treatment Wetlands*; CRC Press: Boca Raton, FL, USA, 2008.
13. Akratos, C.S.; Tsihrintzis, V.A. Effect of temperature, HRT, vegetation and porous media on removal efficiency of pilot-scale horizontal subsurface flow constructed wetlands. *Ecol. Eng.* **2007**, *29*, 173–191. [CrossRef]
14. Papadopoulos, F.H.; Tsihrintzis, V.A. Assessment of a Full-Scale Duckweed Pond System for Septage Treatment. *Environ. Technol.* **2011**, *32*, 795–804. [CrossRef] [PubMed]

15. Kotti, I.P.; Sylaios, G.K.; Tsihrintzis, V.A. Fuzzy modeling for nitrogen and phosphorus removal estimation in free-water surface constructed wetlands. *Environ. Process.* **2016**, *3*, 65–79. [CrossRef]
16. Rezania, S.; Taib, S.M.; Din, M.F.M.; Dahalan, F.A.; Kamyab, H. Comprehensive review on phytotechnology: Heavy metals removal by diverse aquatic plants species from wastewater. *J. Hazard. Mater.* **2016**, *318*, 587–599. [CrossRef] [PubMed]
17. Tsihrintzis, V.A. The Use of Vertical Flow Constructed Wetlands in Wastewater Treatment. *Water Resour. Manag.* **2017**, *31*, 3245–3270. [CrossRef]
18. Zhou, X.; He, Z.L.; Jones, K.D.; Li, L.; Stoffella, P.J. Dominating aquatic macrophytes for the removal of nutrients from waterways of the Indian River Lagoon basin, South Florida, USA. *Ecol. Eng.* **2017**, *101*, 107–119. [CrossRef]
19. Papadopoulos, N.; Zalidis, G. The use of *Typha Latifolia*, L. in constructed wetland microcosms for the remediation of herbicide terbuthylazine. *Environ. Process.* **2019**, *6*, 985–1003. [CrossRef]
20. Wang, J.; Wang, W.; Xiong, J.; Li, L.; Zhao, B.; Sohail, I.; He, Z. A constructed wetland system with aquatic macrophytes for cleaning contaminated runoff/storm water from urban area in Florida. *J. Environ. Manag.* **2021**, *280*, 111794. [CrossRef]
21. Kotti, E.P.; Gikas, G.D.; Tsihrintzis, V.A. Effect of Temperature, HRT, Vegetation and Porous Media on Removal Efficiency of Pilot-Scale Free Water Surface Flow Constructed Wetlands. *Ecol. Eng.* **2010**, *36*, 862–875. [CrossRef]
22. Papadopoulos, F.H.; Tsihrintzis, V.A.; Zdragas, A.G. Removal of Faecal Bacteria from Septage by Treating it in a Full-scale Duckweed-covered Pond System. *J. Environ. Manag.* **2011**, *92*, 3130–3135. [CrossRef]
23. Stefanakis, A.I.; Tsihrintzis, V.A. Effect of Loading, Resting Period, Temperature, Porous Media, Vegetation and Aeration on Performance of Pilot-scale Vertical Flow Constructed Wetlands. *J. Chem. Eng.* **2012**, *181–182*, 416–430. [CrossRef]
24. Stefanakis, A.I.; Tsihrintzis, V.A. Heavy Metal Fate in Pilot-scale Sludge Drying Reed Beds under Various Design and Operation Conditions. *J. Hazard. Mater.* **2012**, *213–214*, 393–405. [CrossRef] [PubMed]
25. Papaevangelou, V.A.; Gikas, G.D.; Tsihrintzis, V.A.; Antonopoulou, M.; Konstantinou, I.K. Removal of Endocrine Disrupting Chemicals in HSF and VF Pilot-scale Constructed Wetlands. *Chem. Eng. J.* **2016**, *294*, 146–156. [CrossRef]
26. Gikas, G.D.; Pérez-Villanueva, M.; Tsioras, M.; Alexoudis, C.; Pérez-Rojas, G.; Masís-Mora, M.; Lizano-Fallas, V.; Rodríguez-Rodríguez, C.E.; Vryzas, Z.; Tsihrintzis, V.A. Low-cost Approaches for the Removal of Terbuthylazine from Agricultural Wastewater: Constructed Wetlands and Biopurification System. *Chem. Eng. J.* **2017**, *335*, 647–656. [CrossRef]
27. Papaevangelou, V.A.; Gikas, G.D.; Tsihrintzis, V.A. Chromium Removal from Wastewater Using HSF and VF Pilot-Scale Constructed Wetlands: Overall Performance, and Fate and Distribution of this Element within the Wetland Environment. *Chemosphere* **2017**, *168*, 716–730. [CrossRef]
28. Papaevangelou, V.A.; Gikas, G.D.; Vryzas, Z.; Tsihrintzis, V.A. Treatment of agricultural equipment rinsing water containing a fungicide in pilot-scale horizontal subsurface flow constructed wetland. *Ecol. Eng.* **2017**, *101*, 193–200. [CrossRef]
29. Gikas, G.D.; Vryzas, Z.; Tsihrintzis, V.A. S-metolachlor Herbicide Removal in Pilot-scale Horizontal Subsurface Flow Constructed Wetlands. *Chem. Eng. J.* **2018**, *339*, 108–116. [CrossRef]
30. Gikas, G.D.; Papaevangelou, V.A.; Tsihrintzis, V.A.; Antonopoulou, M.; Konstantinou, I.K. Removal of emerging pollutants in HSF and VF pilot-scale constructed wetlands. *Processes* **2021**, *9*, 2200. [CrossRef]
31. Gikas, G.D.; Vryzas, Z.; Karametos, I.; Tsihrintzis, V.A. Pyraclostrobin removal in pilot-scale horizontal subsurface flow constructed wetlands and in porous media filters. *Processes* **2022**, *10*, 414. [CrossRef]
32. Díaz, F.J.; Anthony, T.O.; Dahlgren, R.A. Agricultural pollutant removal by constructed wetlands: Implications for water management and design. *Agric. Water Manag.* **2012**, *104*, 171–183. [CrossRef]
33. Ioannidou, V.G.; Pearson, J.M. Hydraulic and design parameters in full-scale constructed wetlands and treatment units: Six case studies. *Environ. Process.* **2018**, *5*, 5–22. [CrossRef]
34. Srivastava, J.; Gupta, A.; Chandra, H. Managing water quality with aquatic macrophytes. *Rev. Environ. Sci. Biotechnol.* **2008**, *7*, 255–266. [CrossRef]
35. Priya, A.; Avishek, K.; Pathak, G. Assessing the potentials of *Lemna minor* in the treatment of domestic wastewater at pilot scale. *Environ. Monit. Assess.* **2012**, *184*, 4301–4307. [CrossRef] [PubMed]
36. Abdul Aziz, N.I.H.; Mohd Hanafiah, M.; Halim, N.H.; Fidri, P.A.S. Phytoremediation of TSS, NH_3-N and COD from Sewage Wastewater by *Lemna minor L.; Salvinia minima, Ipomea aquatica* and *Centella asiatica*. *Appl. Sci.* **2020**, *10*, 5397. [CrossRef]
37. Sarkheil, M.; Safari, O. Phytoremediation of nutrients from water by aquatic floating duckweed (*Lemna minor*) in rearing of African cichlid (*Labidochromis lividus*) fingerlings. *Environ. Technol. Innov.* **2020**, *18*, 100747. [CrossRef]
38. Liu, C.; Dai, Z.; Sun, H. Potential of duckweed (*Lemna minor*) for removal of nitrogen and phosphorus from water under salt stress. *J. Environ. Manag.* **2017**, *187*, 497–503. [CrossRef]
39. Ceschin, S.; Sgambato, V.; Ellwood, N.T.W.; Zuccarello, V. Phytoremediation performance of Lemna communities in a constructed wetland system for wastewater treatment. *Environ. Exp. Bot.* **2019**, *162*, 67–71. [CrossRef]
40. Kamyab, H.; Chelliapan, S.; Din, M.F.M.; Shahbazian-Yassar, R.; Rezania, S.; Khademi, T.; Kumar, A.; Azimi, M. Evaluation of *Lemna minor* and *Chlamydomonas* to treat palm oil mill effluent and fertilizer production. *J. Water Process. Eng.* **2017**, *17*, 229–236. [CrossRef]
41. Sudiarto, S.I.A.; Renggaman, A.; Choi, H.L. Floating aquatic plants for total nitrogen and phosphorus removal from treated swine wastewater and their biomass characteristics. *J. Environ. Manag.* **2019**, *231*, 763–769. [CrossRef]
42. Dosnon-Olette, R.; Couderchet, M.; Eullaffroy, P. Phytoremediation of fungicides by aquatic macrophytes: Toxicity and removal rate. *Ecotoxicol. Environ. Saf.* **2009**, *72*, 2096–2101. [CrossRef]

43. Dosnon-Olette, R.; Couderchet, M.; Biagianti, S.; Eullaffroy, P. Toxicity and removal of pesticides by selected aquatic plants. *Chemosphere* **2008**, *70*, 1414–1421. [CrossRef]
44. Dosnon-Olette, R.; Couderchet, M.; El Arfaoui, A.; Sayen, S.; Eullaffroy, P. Influence of initial pesticide concentrations and plant population density on dimethomorph toxicity and removal by two duckweed species. *Sci. Total Environ.* **2010**, *408*, 2254–2259. [CrossRef] [PubMed]
45. Prasertsup, P.; Ariyakanon, N. Removal of chlorpyrifos by water lettuce (*Pistia stratiotes* L.) and duckweed (*Lemna minor* L.). *Int. J. Phytoremediation* **2011**, *13*, 383–395. [CrossRef] [PubMed]
46. Dosnon-Olette, R.; Couderchet, M.; Oturan, M.A.; Oturan, N.; Eullaffroy, P. Potential use of *Lemna minor* for the phytoremediation of isoproturon and glyphosate. *Int. J. Phytoremediation* **2011**, *13*, 601–612. [CrossRef] [PubMed]
47. Rice, P.J.; Anderson, T.A.; Coats, J.R. Phytoremediation of herbicide-contaminated surface water with aquatic plants. *ACS Publ.* **1997**, *10*, 133–151. [CrossRef]
48. Mitsou, K.; Koulianou, A.; Lambropoulou, D.; Pappas, P.; Albanis, T.; Lekka, M. Growth rate effects, responses of antioxidant enzymes and metabolic fate of the herbicide Propanil in the aquatic plant *Lemna minor*. *Chemosphere* **2006**, *62*, 275–284. [CrossRef] [PubMed]
49. Gatidou, G.; Stasinakis, A.S.; Iatrou, E.I. Assessing single and joint toxicity of three phenylurea herbicides using *Lemna minor* and Vibrio fischeri bioassays. *Chemosphere* **2015**, *119*, S69–S74. [CrossRef] [PubMed]
50. Wang, F.; Liu, D.; Qu, H.; Chen, L.; Zhou, Z.; Wang, P. A full evaluation for the enantiomeric impacts of lactofen and its metabolites on aquatic macrophyte *Lemna minor*. *Water Res.* **2016**, *101*, 55–63. [CrossRef]
51. Tagun, R.; Boxall, A.B. The Response of *Lemna minor* to Mixtures of Pesticides that are Commonly used in Thailand. *Bull. Environ. Contam Toxicol.* **2018**, *100*, 516–523. [CrossRef]
52. Kostopoulou, S.; Ntatsi, G.; Arapis, G.; Aliferis, K.A. Assessment of the effects of metribuzin, glyphosate, and their mixtures on the metabolism of the model plant *Lemna minor* L. applying metabolomics. *Chemosphere* **2020**, *239*, 124582. [CrossRef]
53. Fox, L.J.; Struik, P.C.; Appleton, B.L.; Rule, J.H. Nitrogen phytoremediation by water hyacinth (*Eichhornia crassipes (Mart.) Solms*). *Water Air Soil Pollut.* **2008**, *194*, 199–207. [CrossRef]
54. Nabi, A.; Alam, A.K.M.R.; Hoque, S. Treatment of wastewater with free floating aquatic macrophyte—*Eichhornia crassipes*. *Jahangirnagar Univ. Environ. Bull* **2016**, *5*, 1–9.
55. Osti, J.A.S.; do Carmo, C.F.; Cerqueira, M.A.S.; Giamas, M.T.D.; Peixoto, A.C.; Vaz-dos-Santos, A.M.; Mercante, C.T.J. Nitrogen and phosphorus removal from fish farming effluents using artificial floating islands colonized by *Eichhornia crassipes*. *Aquac. Rep.* **2020**, *17*, 100324. [CrossRef]
56. Kutty, S.R.M.; Ngatenah, S.N.I.; Isa, M.H.; Malakahmad, A. Nutrients removal from municipal wastewater treatment plant effluent using *Eichhornia crassipes*. *World Acad. Eng. Technol.* **2009**, *60*, 826–831.
57. Sooknah, R.D.; Wilkie, A.C. Nutrient removal by floating aquatic macrophytes cultured in anaerobically digested flushed dairy manure wastewater. *Ecol. Eng.* **2004**, *22*, 27–42. [CrossRef]
58. Chen, X.; Chen, X.; Wan, X.; Weng, B.; Huang, Q. Water hyacinth (*Eichhornia crassipes*) waste as an adsorbent for phosphorus removal from swine wastewater. *Bioresour. Technol.* **2010**, *101*, 9025–9030. [CrossRef] [PubMed]
59. Aremu, A.S.; Ojoawo, S.O.; Alade, G.A. Water hyacinth (*Eichhornia crassipes*) culture in sewage: Nutrient removal and potential applications of bye-products. *Transnatl. J. Sci. Technol.* **2012**, *2*, 103–114.
60. Zhao, F.; Xi, S.; Yang, X.; Yang, W.; Li, J.; Gu, B.; He, Z. Purifying eutrophic river waters with integrated floating island systems. *Ecol. Eng.* **2012**, *40*, 53–60. [CrossRef]
61. Wang, Z.; Zhang, Z.; Zhang, Y.; Zhang, J.; Yan, S.; Guo, J. Nitrogen removal from Lake Caohai, a typical ultra-eutrophic lake in China with large scale confined growth of *Eichhornia crassipes*. *Chemosphere* **2013**, *92*, 177–183. [CrossRef]
62. Lima, M.X.; Carvalho, K.Q.; Passig, F.H.; Borges, A.C.; Filippe, T.C.; Azevedo, J.C.R.; Nagalli, A. Performance of different substrates in constructed wetlands planted with *E. crassipes* treating low-strength sewage under subtropical conditions. *Sci. Total Environ.* **2018**, *630*, 1365–1373. [CrossRef]
63. Xia, H.; Ma, X. Phytoremediation of ethion by water hyacinth (*Eichhornia crassipes*) from water. *Bioresour. Technol.* **2006**, *97*, 1050–1054. [CrossRef]
64. Anudechakul, C.; Vangnai, A.S.; Ariyakanon, N. Removal of chlorpyrifos by water hyacinth (*Eichhornia crassipes*) and the role of a plant-associated bacterium. *Int. J. Phytoremediation* **2015**, *17*, 678–685. [CrossRef] [PubMed]
65. Alencar, B.T.B.; Ribeiro, V.H.V.; Cabral, C.M.; dos Santos, N.M.C.; Ferreira, E.A.; Francino, D.M.T.; dos Santos, J.B.; Silva, D.V.; de Freitas Souza, M. Use of macrophytes to reduce the contamination of water resources by pesticides. *Ecol. Indic.* **2020**, *109*, 105785. [CrossRef]
66. Akinbile, C.O.; Ikuomola, B.T.; Olanrewaju, O.O.; Babalola, T.E. Assessing the efficacy of *Azolla pinnata* in four different wastewater treatment for agricultural re-use: A case history. *Sustain. Water Resour. Manag.* **2019**, *5*, 1009–1015. [CrossRef]
67. Soman, D.; Anitha, V.; Arora, A. Fitoremediation of municipal sewage water with *Azolla microphylla*. *Int. J. Adv. Res.* **2018**, *6*, 101–108. [CrossRef] [PubMed]
68. Muvea, F.M.; Ogendi, G.M.; Omondi, S.O. Nutrient removal efficiency by floating macrophytes; *Lemna minor* and *Azolla pinnata* in a constructed wetland. *Glob. J. Environ. Sci. Manag.* **2019**, *5*, 415–430.
69. Golzary, A.; Tavakoli, O.; Rezaei, Y.; Karbassi, A. Wastewater treatment by *Azolla Filiculoides*: A study on color, odor, COD, nitrate, and phosphate removal. *Pollution* **2018**, *4*, 69–76.

70. European Food Safety Authority (EFSA). Conclusion on pesticide peer review regarding the risk assessment of the active substance myclobutanil. *EFSA J.* **2009**, *7*, 298r. [CrossRef]
71. European Food Safety Authority (EFSA). Conclusion regarding the peer review of the pesticide risk assessment of the active substance imidacloprid. *EFSA J.* **2008**, *6*, 148r.
72. European Food Safety Authority (EFSA). Conclusion on the peer review of the pesticide risk assessment of the active substance difenoconazole. *EFSA J.* **2011**, *9*, 1967. [CrossRef]
73. European Food Safety Authority (EFSA). Peer review of the pesticide risk assessment of the active substance thiacloprid. *EFSA J.* **2019**, *17*, e05595.
74. European Food Safety Authority (EFSA). Conclusion on the peer review of the pesticide risk assessment of the active substance dimethomorph. *EFSA Sci. Rep.* **2006**, *82*, 1–69.
75. Penman, H.L. Natural evaporation from open water, bare soil and grass. In *Proceedings of the Royal Society of London*; Series A: Mathematical and Physical Sciences; The Royal Society: London, UK, 1948; Volume 193, pp. 120–145.
76. Monteith, J.L. Evaporation and environment. In *Symposia of the Society for Experimental Biology*; Cambridge University Press (CUP): Cambridge, UK, 1965; Volume 19, pp. 205–234.
77. Ozengin, N.; Elmaci, A. Performance of Duckweed (*Lemna minor* L.) on different types of wastewater treatment. *J. Environ. Biol.* **2007**, *28*, 307–314. [PubMed]
78. Wang, H.; Zhang, H.; Cai, G. An application of phytoremediation to river pollution remediation. *Procedia Environ. Sci.* **2011**, *10*, 1904–1907. [CrossRef]
79. Qin, H.; Zhang, Z.; Liu, M.; Liu, H.; Wang, Y.; Wen, X.; Zhang, Y.; Yan, S. Site test of phytoremediation of an open pond contaminated with domestic sewage using water hyacinth and water lettuce. *Ecol. Eng.* **2016**, *95*, 753–762. [CrossRef]
80. Kumari, M.; Tripathi, B.D. Effect of aeration and mixed culture of *Eichhornia crassipes* and *Salvinia natans* on removal of wastewater pollutants. *Ecol. Eng.* **2014**, *62*, 48–53. [CrossRef]
81. Xu, L.; Cheng, S.; Zhuang, P.; Xie, D.; Li, S.; Liu, D.; Li, Z.; Wang, F.; Xing, F. Assessment of the nutrient removal potential of floating native and exotic aquatic macrophytes cultured in swine manure wastewater. *Int. J. Environ. Res. Public Health* **2020**, *17*, 1103. [CrossRef]
82. Ekperusi, A.O.; Sikoki, F.D.; Nwachukwu, E.O. Application of common duckweed (*Lemna minor*) in phytoremediation of chemicals in the environment: State and future perspective. *Chemosphere* **2019**, *223*, 285–309. [CrossRef] [PubMed]
83. Nie, J.; Sun, Y.; Zhou, Y.; Kumar, M.; Usman, M.; Li, J.; Shao, J.; Wang, L.; Tsang, T.C.W. Bioremediation of water containing pesticides by microalgae: Mechanisms, methods, and prospects for future research. *Sci. Total Environ.* **2020**, *707*, 136080. [CrossRef] [PubMed]
84. Gondi, R.; Kavitha, S.; Kannah, R.Y.; Karthikeyan, O.P.; Kumar, G.; Tyagi, V.K.; Banu, J.R. Algal-based system for removal of emerging pollutants from wastewater: A review. *Bioresour. Technol.* **2022**, *344*, 126245. [CrossRef]
85. Friesen-Pankratz, B.B.; Doebel, C.C.; Farenhorst, A.A.; Gordon Goldsborough, L. Interactions between algae (*Selenastrum capricornutum*) and pesticides: Implications for managing constructed wetlands for pesticide removal. *J. Environ. Sci. Health Part B* **2003**, *38*, 147–155. [CrossRef]
86. Avila, R.; Peris, A.; Eljarrat, E.; Vicent, T.; Blánquez, P. Biodegradation of hydrophobic pesticides by microalgae: Transformation products and impact on algae biochemical methane potential. *Sci. Total Environ.* **2021**, *754*, 142114. [CrossRef] [PubMed]
87. Riaz, G.; Tabinda, A.B.; Iqbal, S.; Yasar, A.; Abbas, M.; Khan, A.M.; Mahfooz, Y.; Baqar, M. Phytoremediation of organochlorine and pyrethroid pesticides by aquatic macrophytes and algae in freshwater systems. *Int. J. Phytoremediation* **2017**, *19*, 894–898. [CrossRef] [PubMed]

toxics

MDPI

Article

Selected Biochemical Markers Change after Oral Administration of Pesticide Mixtures in Honey Bees

Pawel Migdal [1,*], Agnieszka Murawska [1], Ewelina Berbeć [1], Mateusz Plotnik [1], Anita Skorus [1] and Krzysztof Latarowski [2]

1 Department of Environment, Hygiene and Animal Welfare, Bee Division, The Wroclaw University of Environmental and Life Sciences, 25 C.K. Norwida St., 51-630 Wroclaw, Poland
2 Department of Human Nutrition, The Wroclaw University of Environmental and Life Sciences, 25 C.K. Norwida St., 51-630 Wroclaw, Poland
* Correspondence: pawel.migdal@upwr.edu.pl

Abstract: The honey bee is an important pollinator. In the environment, it can be exposed to many harmful factors, such as pesticides. Nowadays, attention is paid to evaluating the potentially harmful effects of these substances. This study aimed to evaluate the effect of worst-case environmental concentrations of pesticide mixtures on honey bee survival and selected physiological markers (the activity of ALT, AST, ALP, and GGTP, and the concentration of albumin, creatinine, urea, and uric acid). Pesticides of three different groups (insecticide—acetamiprid, herbicide—glyphosate, and fungicide—tebuconazole) and their mixtures were resolved in 50% (*w*/*v*) sucrose solution and given to bees *ad libitum*. After 24 h, hemolymph was collected. All mixtures caused higher mortality than single pesticides. Pesticides in mixtures caused disturbances in biochemical markers, and in some cases the interaction between pesticides was synergistic. The mixtures had individual effects on physiology, and the results were sensitive to changes in proportions.

Keywords: honey bee; *Apis mellifera*; pesticide compositions; insect physiology; biochemical markers

1. Introduction

As a pollinator, the honey bee has a positive effect on increasing agricultural yields and preserving biodiversity. Bee pollination is valued at around 15 billion USD in the US, 19 billion USD in Europe, and 69 billion USD in East Asia [1]. For many years, the toxicity of pesticides to these insects has drawn the attention of researchers. One of the criteria for the division of pesticides concerns the type of target pest [2]. This classification includes, among others, insecticides, fungicides, and herbicides [3,4]. Honey bee contact with the pesticide at a sub-lethal dose may affect their behavior and/or physiology. Poisoning may occur after contact and/or oral exposure. Systemic pesticides, commonly used in developed countries, spread in plant tissues and may accumulate in plant nectar and pollen. In addition to exposure to harmful substances in the environment, bees may also come into contact with them in the hive, as they collect potentially contaminated nectar or pollen and store it in combs [5]. Due to the fact that many different pesticides are used in plant production, and their residues can accumulate in the environment, bees come into contact with many different toxins in various concentrations and proportions simultaneously [6]. When combined, an additive effect (i.e., the sum of the individual substances), synergistic effect (greater than an additive effect), or antagonistic effect (less than an additive effect) may occur. According to Regulation (EC) No 1107/2009 of the European Parliament and of the Council of 21 October 2009 concerning the placing of plant protection products on the market and repealing Council Directives 79/117/EEC and 91/414/EEC, mixing pesticides is a legal action unless it is expressly prohibited on the label of the product.

So far, the exposure of honey bees to pesticides has been shown to affect bee motor activity, navigation, feeding, learning ability, and memory, to weaken the immune and reproductive systems, and to activate the body's antioxidant and detoxification mechanisms [7–12]. In order to fight harmful substances, the honey bee organism has developed many defense mechanisms within their detoxification and antioxidant systems. The detoxification mechanisms of the bee body mainly include enzymes involved in the metabolism of toxins or the detoxification process, i.e., cytochrome P450 monooxygenase (P450), glutathione transferase (GST), carboxylesterase (COE), aspartate aminotransferase (AST), alkaline aminotransferase (ALT) (ALP), gamma-glutamyl transpeptidase (GGTP), and bilirubin [12,13]. The mechanisms of the antioxidant system are designed to remove free radicals from the body. Antioxidants give electrons to free radicals and, as a result, the possibility of oxidizing other components is blocked. Among antioxidants, enzymatic antioxidants, such as glutathione peroxidase (GPX), catalases (CAT), superoxide dismutase (SOD), and glucose-6-phosphate dehydrogenase (GP6D), can be distinguished from non-enzymatic antioxidants (e.g., albumin, creatinine, glutathione, uric acid, urea, and vitamins) [11,13]. Many studies have confirmed that insecticides change enzyme activity and the content of some key substances [11,12,14–17]. However, reference values have not been estimated and there is still a lack of information about the influence of pesticide mixtures.

Our research aimed to investigate how oral exposure to pesticide mixtures affects the activity of selected hemolymph enzymes and non-enzymatic antioxidants of worker honey bees, i.e., ALT, ALP, AST, GGTP, albumin, creatinine, uric acid, and urea.

2. Materials and Methods

2.1. Research Material—Honey Bee Workers

Honey bee (*Apis mellifera* carnica) colonies used for research were treated against Varroa destructor using amitraz fumigation four times at 4-day intervals (12.5 mg/tablet; Apiwarol®, Biowet, Pulawy, Poland) before starting the experiment. To monitor the number of Nosema spp. spores, the hemocytometer method was used (30 bees per hive in three repetitions). After 28 days from the last fumigation, we selected 3 frames with bee brood in 20 days of apian development. Next, we took the brood to the laboratory and incubated it at temperatures of 34 and 70% of relative humidity. After 24 h, we collected the bees. Worker bees were placed in wooden cages (20 cm × 15 cm × 7 cm), each containing 100 workers and two inner feeders with 50% (w/v) sucrose solution (Chempur®, Piekary Śląskie, Poland) ad libitum. The adaptation process lasted 24 h at a temperature of 25 °C ± 0.5 °C and relative humidity of 70% ± 5%. Caged bees were maintained in the incubator in the same conditions described above until being used for the experiment [18]. Two-day-old bees were used in the study, and the bees were divided into 14 groups. Dead bees were utilized by a special biohazard waste company.

2.2. Experimental Setup

Each group consisted of ten cages. The experiment was performed by feeding bees 50% (w/v) sucrose solution containing the established concentrations of particular pesticides for 24 h. Bees in the experimental groups were exposed to the single-pesticide commercial formulations or their trinary mixtures in different proportions. The control group was fed an untreated 50% (w/v) sucrose solution. The experiment used doses of pesticides recommended by the manufacturer for the selected active substance, which represented worst-case environmental concentrations. The bees were exposed to the insecticide Mospilan® 20SP, Target, Kartoszyno, Poland (ai acetamiprid 20%), the herbicide Agrosar® 360 SL, CIECH Sarzyna, Nowa Sarzyna, Poland (ai glyphosate 36%), and the fungicide Tebu® EW, HELM, Hamburg, Germany (ai tebuconazole 25.8%) (Figure 1). Acetamiprid, as a cyano-substituted neonicotinoid, has a higher dose value that kills 50% of honey bees (LD50) compared to the nitro-substituted neonicotinoids (imidacloprid, clothianidin, thiamethoxam) and is considered less toxic than them [5,19]. However, its toxicity can be higher after being mixed with fungicide [20]. The LD_{50} of acetamiprid is 14,530 ng/bee (food exposure) and 8090 ng/bee (contact exposure) [5]. Mortality and syrup intake were recorded after 24 h.

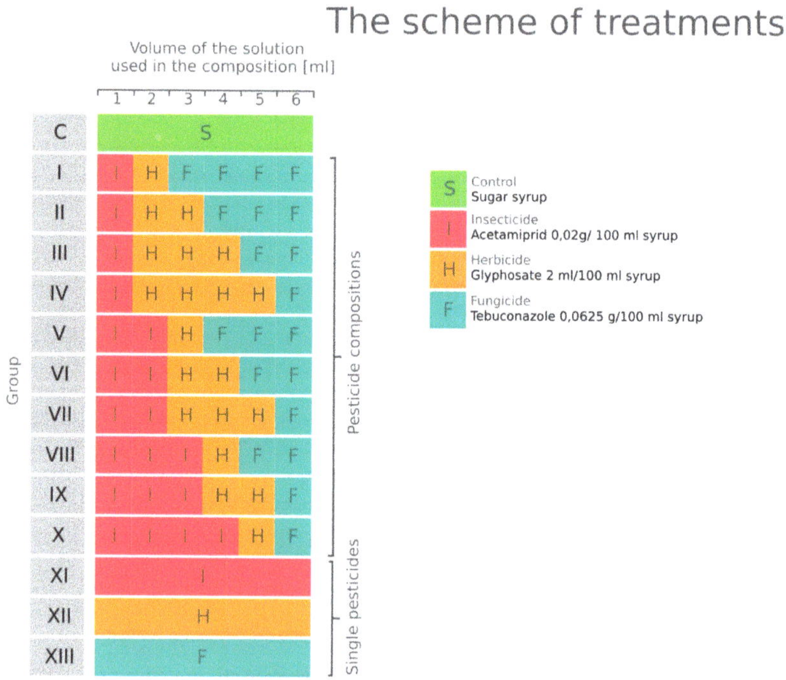

Figure 1. The scheme of treatments. Each group was fed *ad libitum* with a solution of a total volume of 6 mL. The solution composition differed between groups. The group abbreviated as C was a control group, fed with 50% (w/v) sucrose solution. Groups XI–XIII were fed with single pesticides dissolved in 50% (w/v) sucrose solution. Groups I-X were fed with pesticide compositions consisting of each type of pesticide in varying proportions.

2.3. Collection of Hemolymph

Hemolymph was taken from 100 alive worker honey bees from each group immediately after 24 h of oral exposure by removing the antennae with sterile tweezers. Hemolymph was conserved in a 20 µL end-to-end glass capillary without anticoagulant. Hemolymph from the control honey bees was collected at the same time. The test tubes were placed on the cooling block during the operation. After collecting the hemolymph, samples were stored at −80 °C [21].

2.4. Biochemical Analysis

Hemolymph biochemical parameters were determined using the Pentra 400 automated biochemical analyzer by Horiba ABX (Longjumeau, France). The colorimetric method with the use of bromocresol green and the creatinine kinetic method with alkaline picrate were used for measurement of complex formation coloring; urea—the UV enzyme test; uric acid—the enzymatic method using the Trinder reaction with Horiba ABX reagents were used to assess the level of albumin activity. Reagents from Randox (Crumlin, Great Britain) were used to assess the enzymatic activity (aspartate aminotransferase—AST, alkaline phosphatase—ALP, alanine aminotransferase—ALT, gamma-glutamyl transpeptidase—GGTP). Analysis kits for ALP—the photometric kinetic test in accordance with the recommendations of the International Federation of Clinical Chemistry (IFCC); for ALT and AST—the enzymatic method (UV detection) in accordance with IFCC recommendations; and for GGTP—the kinetic photometric test.

2.5. *Statistical Analysis*

Statistical analyses were performed using the R program, version 3.4.4 (R Core Team, 2018, R-3.4.4 for Windows, CRAN, Vienna, Austria), with the RStudio overlay, using, inter alia, the packages "dplyr", "tidyr", "agricolae", and "ggplot2" [22]. The normality of the distribution was checked by the Shapiro–Wilk test, and the differences between the groups by the Kruskal–Wallis test with holm correction for multiple comparisons, $\alpha = 0.05$. In Inkscape, the experiment scheme was made, and the charts were visually improved.

3. Results

3.1. *Mortality and Syrup Intake*

In the control group, there was 0% mortality. Among the pesticides administered separately, none caused acute toxicity, with the highest mortality after fungicide administration being only 3.2% (Figure 2 and Table 1). All compositions showed mortality significantly higher than the control, while all except groups III and IV caused mortality higher than single pesticides. The highest mortality—over 70%—occurred in groups VIII, IX, and IV. Considering syrup intake within separately administered pesticides, the herbicide caused an intake similar to the control, while the fungicide intake was almost 50% lower, but the difference was not statistically significant. Mean values of syrup intake were lower in all compositions compared to the control, but the difference was statistically significant only in group I, with the lowest value over 3× lower than the control.

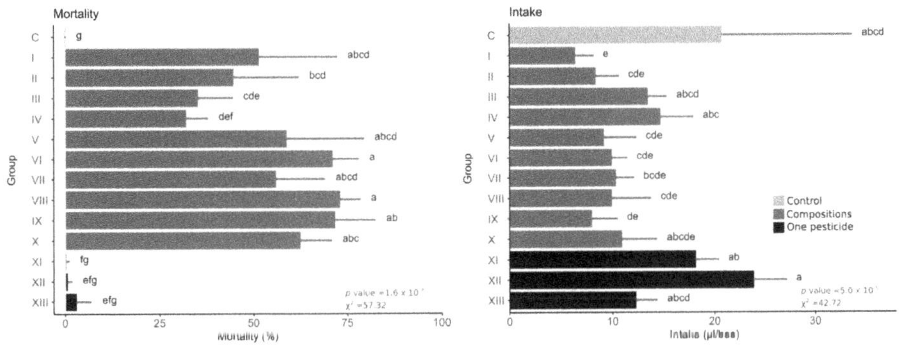

Figure 2. Mortality and syrup intake during 24 h of the experiment. Bars represent the mean, and error bars represent the standard deviation. The same letters between groups within one plot means no statistically significant differences (Kruskal–Wallis test with holm correction for multiple comparisons, $\alpha = 0.05$); the statistical values are shown in the lower right corner of each graph.

Table 1. The dose of each active ingredient per individual bee used in the research.

Group	Acetamiprid (µg)	Glyphosate (µL)	Tebuconazole (µg)
C	0.00	0.00	0.00
I	0.21	0.02	2.67
II	0.28	0.06	2.63
III	0.45	0.14	2.83
IV	0.49	0.20	1.54
V	0.61	0.03	2.88
VI	0.67	0.07	2.08
VII	0.69	0.10	1.08
VIII	1.00	0.03	2.08
IX	0.80	0.05	0.83

Table 1. *Cont.*

Group	Acetamiprid (µg)	Glyphosate (µL)	Tebuconazole (µg)
X	1.47	0.04	1.15
XI	3.64	0.00	0.00
XII	0.00	0.48	0.00
XIII	0.00	0.00	7.75

3.2. Enzymatic Activity

The overall effect of pesticides was an increase in alanine aminotransferase (ALT). Of the single pesticides, only the insecticide increased ALT (Figure 3). All compositions caused an increase except group IV (with the highest amount of herbicide). The highest value was in group VII—more than 7× higher than the control and more than 2× higher than the single insecticide. Concerning aspartate aminotransferase (AST), the overall effect of pesticides was an increase in this parameter. Of the single pesticides, only the insecticide increased AST. All compositions except group IX caused an increase in AST. The highest increase occurred in group VII—more than 4× higher than the control and almost 2× higher than the insecticide, with a high value also in group X (composition with the highest proportion of insecticide). The overall effect of pesticides on alkaline phosphatase (ALP) was an increase in this parameter. Of the single pesticides, the insecticide and fungicide caused an increase in this parameter, while the herbicide had no effect. Most of the compositions increased ALP—only groups III and VII showed no increase. The highest value was found in group II—more than twice as high as the control, and a high value also in group VI. Generally, pesticides caused an increase or a decrease in gamma-glutamyl transpeptidase (GGTP), but the decreases were not statistically significant. Among the single pesticides, the differences were not significant in any, but the herbicide caused a very low GGTP value, while the fungicide caused an increase in GGTP. Among the compositions, groups IX and III caused a lower level of GGTP than in the control; however, the difference was not statistically significant. The highest value of GGTP occurred in group VI (approx. 4× higher than the control), and a significant increase in this indicator was also observed in groups II and VII.

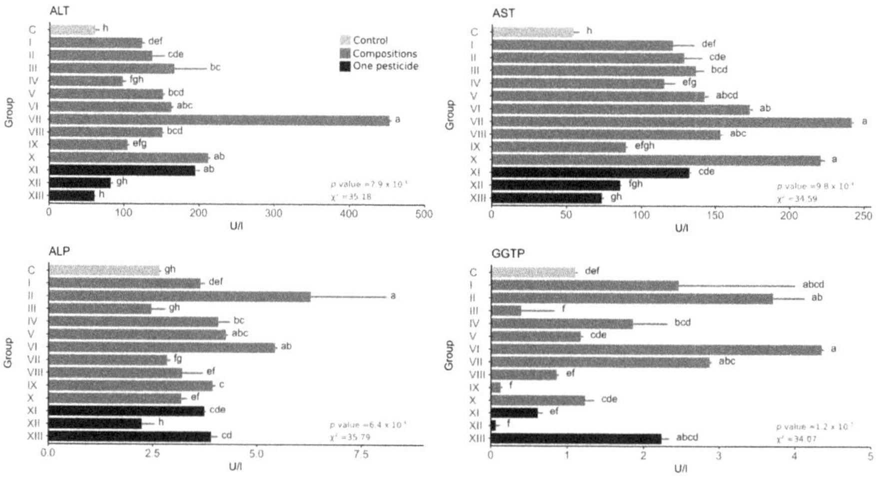

Figure 3. Enzymatic activity. Bars represent the mean, and error bars represent the standard deviation. The same letters between groups within one plot means no statistically significant differences (Kruskal–Wallis test with Holm correction for multiple comparisons, α = 0.05); the statistical values are shown in the lower right corner of each graph.

3.3. *Detoxification System Indicators*

Generally, pesticides caused the albumin levels to rise or fall. Among single pesticides, only the fungicide significantly changed the level of albumin, causing an increase (Figure 4). No composition had a higher level than the fungicide, while the highest values occurred in groups VI, II, and I. In the IX group, the albumin level was 3× lower than in the control. Pesticides generally caused a drop in creatinine levels. Among the pure pesticides, a statistically significant effect occurred only for the insecticide. Of the compositions, only two groups did not cause a statistically significant decrease (I and IV). The lowest value occurred in group IX, while low levels of creatinine also occurred in groups VII, VIII, III, and X. The values of urea acid after pesticide exposure were higher or lower than the control. Neither group differed statistically significantly from the control. The highest values were found in group V (2× higher than the control), and high levels of uric acid were also in groups VI and VIII. The lowest values were in groups I and IX and the single insecticide. Pesticides caused both an increase and a decrease in urea levels, with the higher values not statistically different from the control. The lowest value occurred in group IX (approx. 4× lower than the control), and low urea rates also occurred in groups IV and III.

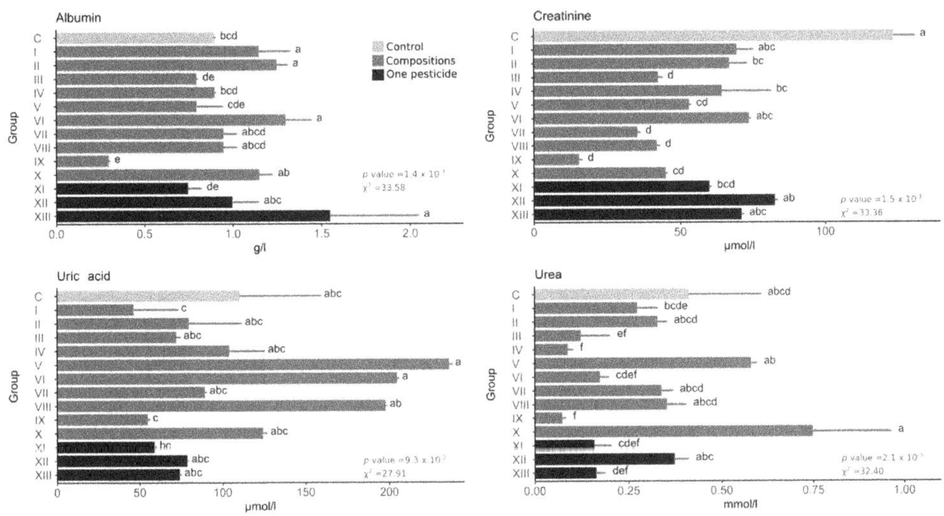

Figure 4. Detoxification system indicators. Bars represent the mean, and error bars represent the standard deviation. The same letters between groups within one plot mean no statistically significant differences (Kruskal–Wallis test with holm correction for multiple comparisons, $\alpha = 0.05$); the statistical values are shown in the lower right corner of each graph.

4. Discussion

The pesticide compositions induced much higher acute toxicity than single pesticides. The differences were so large that this relationship can be called synergistic for all tested compositions. The synergistic mortality-increasing effect of the combinations of different types of pesticides has been previously observed [14,17,23]. This study suggests that the use of Acetamiprid, Glyphosate, and Tebuconazole in mixtures can significantly increase bee mortality. The specified toxicity in the first 24 h of our study was high. A bee in the environment, had it come into contact with such a set of pesticides, would not have returned to the hive. The concentrations of the pesticides did not exceed the manufacturer's recommendations and were given to the bees separately; they did not show a significant effect on survival with short-term 24-h exposure. Walker et al. [24], investigating the linkage of an insecticide, fungicide, and adjuvant, had similar observations. The pesticides in combinations caused higher mortality than those used alone in the studies by

Belsky et al. [25]. This shows that it is very important to compare the honey bee toxicity of individual pesticides and their mixtures. Additionally, the high toxicity of the pesticide compositions was confirmed by the syrup intake generally being lower in groups fed with the pesticide compositions than in groups fed with the single pesticides and control. Higher mortality was observed in the groups with pesticide mixtures despite lower syrup intake.

The pesticide mixtures had an individual effect; for each enzyme tested, there were compositions whose effect was greater than that of single pesticides. The general effect of single pesticides and mixtures on AST, ALT, and ALP was, if one occurred, an increase in these indicators. The observed effect on GGTP was either the increase or decrease of its activity compared to the control. Our research showed statistically significant changes in the activity of ALT, AST, and ALP compared to the control group in the case of most of the mixtures of pesticides. Similar observations of changes in activity were observed in the studies by Zhu et al. [17] when combining two insecticides containing the active substances imidacloprid and oxamyl, which resulted in a decrease in the activity of phenoloxidase—an immunity enzyme—and this effect did not occur when these substances were administered separately. Two binary compositions, thiamethoxam + λ-cyhalothrin and thiamethoxam + abamectin, caused a significant decrease in the activity and expression of a group of key insect detoxifying enzymes—glutathione S-transferases. A significant increase in mortality was also observed in these groups compared to the effects of single pesticides. In addition, thiamethoxam + abamectin caused a significant increase in ALP expression with a simultaneous decrease in the activity of this enzyme [23]. Changes in the activity of AST, ALT, and ALP have also been demonstrated in studies on the effect of imidacloprid [12]. In our research, the effect of selected plant protection products was an increase in the activity of these enzymes. Changes in the activity of these enzymes were also observed when bees were exposed to other substances. Bromfenvinphos, which is a substance used to treat bees during the infestation of the Varroa destructor mite, caused a decrease in the activity of AST, ALT, and ALP [26]. A similar effect was observed with the antifungal antibiotic amphotericin B [27]. During long-term coenzyme Q10 supplementation, an increase in ALT, AST, and ALP activity was observed [28]. Caffeine also caused an increase in the activity of these enzymes, and a similar effect was observed for piperine [29] and curcumin [30]. Increased concentrations of enzymes in the hemolymph may indicate a greater need for them by the organism.

Pesticides and mixtures caused a decrease in creatinine concentration, while in the case of albumin, urea, and uric acid an increase was observed in some groups, while a decrease was observed in others. Some mixtures of pesticides had a greater effect than individual pesticides; thus, in the case of the detoxification system, the individual effect of the mixture was also visible. The single use of the fungicide and some mixtures increased the concentration of albumin, and a similar effect was observed with the administration of bromfenvinphos used in the treatment of varroosis. Bromfenvinphos also caused a decrease in creatinine, urea, and uric acid levels [31]. Long-term administration of coenzyme Q10 caused a decrease in each of these indicators [28]. Caffeine supplementation caused an increase in uric acid and creatinine concentration and a decrease in albumin and urea concentration, and curcumin had a similar effect [30]. Albumin, creatinine, uric acid, and urea are substances that can also be classified as non-enzymatic antioxidants [28], hence they participate in the detoxification of oxidative stress agents.

5. Conclusions

Assessing the degree of pesticide effects on the honey bee in different combinations of substances continuously provides new information. It can be seen that the multidirectional exposure of bees to pesticides contributes to an increase in mortality and the disruption of the activity of biochemical markers. Such disturbances in the functioning of the organism may cause higher sensitivity to external factors. Showing the effects of single substances and comparing them with the effects of their mixtures is the basis for developing this research area.

Author Contributions: Conceptualization, P.M., A.M., and K.L; methodology, P.M. and A.M.; validation, P.M. and E.B.; formal analysis, E.B.; investigation, P.M., A.M., M.P., A.S. and K.L.; resources, P.M., A.M., K.L., A.S. and M.P.; data curation, P.M. and E.B.; writing—original draft preparation, P.M., E.B., K.L., and A.M.; writing—review and editing, P.M., A.M., E.B., K.L., and M.P.; visualization, E.B.; supervision, P.M.; project administration, P.M.; funding acquisition, P.M. and K.L. All authors have read and agreed to the published version of the manuscript.

Funding: This work was co-financed by the Department of Environment, Hygiene and Animal Welfare, project number B010/0005/22, and the Department of Human Nutrition, project number B010/0013/22. This work was supported by the Wroclaw University of Environmental and Life Sciences (Poland) as the Ph.D. research program "Bon doktoranta SD UPWr" number N020/1754/2021 and innovative scientist number N070/0014/21.

Institutional Review Board Statement: Not applicable.

Informed Consent Statement: Not applicable.

Data Availability Statement: The datasets generated and/or analyzed during the current study are available from the corresponding author on reasonable request.

Conflicts of Interest: The authors declare no conflict of interest. The funders had no role in the design of the study; in the collection, analyses, or interpretation of data; in the writing of the manuscript; or in the decision to publish the results.

References

1. Allsopp, M.H.; de Lange, W.J.; Veldtman, R. Valuing Insect Pollination Services with Cost of Replacement. *PLoS ONE* **2008**, *3*, e3128. [CrossRef] [PubMed]
2. Tudi, M.; Daniel Ruan, H.; Wang, L.; Lyu, J.; Sadler, R.; Connell, D.; Chu, C.; Phung, D.T. Agriculture Development, Pesticide Application and Its Impact on the Environment. *Int. J. Environ. Res. Public Health* **2021**, *18*, 1112. [CrossRef] [PubMed]
3. Fishel, F.M.; Ferrell, J.A. *Managing Pesticide Drift*; Agronomy Department. PI232; University of Florida: Gainesville, FL, USA, 2013; Available online: http://edis.ifas.ufl.edu/pi232 (accessed on 1 April 2022).
4. Kaur, R.; Mavi, G.K.; Raghav, S.; Khan, I. Pesticides Classification and its Impact on Environment. *Int. J. Curr. Microbiol. Appl. Sci.* **2019**, *8*, 1889–1897. [CrossRef]
5. Decourtye, A.; Devillers, J. Ecotoxicity of Neonicotinoid Insecticides to Bees. In *Insect Nicotinic Acetylcholine Receptors*; Thany, S.H., Ed.; Advances in Experimental Medicine and Biology; Springer: New York, NY, USA, 2010; Volume 683, pp. 85–95. [CrossRef]
6. Tang, F.H.M.; Lenzen, M.; McBratney, A.; Maggi, F. Risk of pesticide pollution at the global scale. *Nat. Geosci.* **2021**, *14*, 206–210. [CrossRef]
7. Decourtye, A.; Devillers, J.; Cluzeau, S.; Charreton, M.; Pham-Delègue, M.H. Effects of imidacloprid and deltamethrin on associative learning in honeybees under semi-field and laboratory conditions. *Ecotoxicol. Environ. Saf.* **2004**, *57*, 410–419. [CrossRef] [PubMed]
8. Farooqui, T. A potential link among biogenic amines-based pesticides, learning and memory, and colony collapse disorder: A unique hypothesis. *Neurochem. Int.* **2013**, *62*, 122–136. [CrossRef] [PubMed]
9. Grünewald, B.; Siefert, P. Acetylcholine and Its Receptors in Honeybees: Involvement in Development and Impairments by Neonicotinoids. *Insects* **2019**, *10*, 420. [CrossRef] [PubMed]
10. Migdał, P.; Roman, A.; Popiela-Pleban, E.; Kowalska-Góralska, M.; Opaliński, S. The Impact of Selected Pesticides on Honey Bees. *Pol. J. Environ. Stud.* **2018**, *27*, 787–792. [CrossRef]
11. Paleolog, J.; Wilde, J.; Miszczak, A.; Gancarz, M.; Strachecka, A. Antioxidation Defenses of Apis mellifera Queens and Workers Respond to Imidacloprid in Different Age-Dependent Ways: Old Queens Are Resistant, Foragers Are Not. *Animals* **2021**, *11*, 1246. [CrossRef] [PubMed]
12. Paleolog, J.; Wilde, J.; Siuda, M.; Bąk, B.; Wójcik, Ł.; Strachecka, A. Imidacloprid markedly affects hemolymph proteolysis, biomarkers, DNA global methylation, and the cuticle proteolytic layer in western honeybees. *Apidologie* **2020**, *51*, 620–630. [CrossRef]
13. Łoś, A.; Strachecka, A. Fast and Cost-Effective Biochemical Spectrophotometric Analysis of Solution of Insect "Blood" and Body Surface Elution. *Sensors* **2018**, *18*, 1494. [CrossRef] [PubMed]
14. Almasri, H.; Tavares, D.A.; Pioz, M.; Sené, D.; Tchamitchian, S.; Cousin, M.; Brunet, J.L.; Belzunces, L.P. Mixtures of an insecticide, a fungicide and a herbicide induce high toxicities and systemic physiological disturbances in winter Apis mellifera honey bees. *Ecotoxicol. Environ. Saf.* **2020**, *203*, 111013. [CrossRef] [PubMed]
15. Li, Z.; Li, M.; He, J.; Zhao, X.; Chaimanee, V.; Huang, W.F.; Nie, H.; Zhao, Y.; Su, S. Differential physiological effects of neonicotinoid insecticides on honey bees: A comparison between *Apis mellifera* and *Apis cerana*. *Pestic. Biochem. Physiol.* **2017**, *140*, 1–8. [CrossRef] [PubMed]

16. Wilde, J.; Frączek, R.J.; Siuda, M.; Bąk, B.; Hatjina, F.; Miszczak, A. The influence of sublethal doses of imidacloprid on protein content and proteolytic activity in honey bees (*Apis mellifera* L.). *J. Apic. Res.* **2016**, *55*, 212–220. [CrossRef]
17. Zhu, Y.C.; Yao, J.; Adamczyk, J.; Luttrell, R. Synergistic toxicity and physiological impact of imidacloprid alone and binary mixtures with seven representative pesticides on honey bee (*Apis mellifera*). *PLoS ONE* **2017**, *12*, e0176837. [CrossRef]
18. Medrzycki, P.; Giffard, H.; Aupinel, P.; Belzunces, L.; Chauzat, M.P.; Claßen, C.; Colin, M.E.; Dupont, T.; Girolami, V.; Johnson, R.; et al. Standard methods for toxicology research in Apis mellifera. *J. Apic. Res.* **2013**, *52*, 1–60. [CrossRef]
19. Iwasa, T.; Motoyama, N.; Ambrose, J.T.; Roe, R.M. Mechanism for the differential toxicity of neonicotinoid insecticides in the honey bee, Apis mellifera. *Crop Prot.* **2004**, *23*, 371–378. [CrossRef]
20. Brunet, J.L.; Badiou, A.; Belzunces, L.P. In vivo metabolic fate of [14C]-acetamiprid in six biological compartments of the honeybee, *Apis mellifera* L. *Pest Manag. Sci.* **2005**, *61*, 742–748. [CrossRef] [PubMed]
21. Migdał, P.; Murawska, A.; Roman, A. A modified standardized method to extract and store insect hemolymph with use of a glass capillary. *J. Apic. Sci.* **2020**, *64*, 165–168. [CrossRef]
22. R Core Team. *R: A Language and Environment for Statistical Computing*; R Foundation for Statistical Computing: Vienna, Austria, 2018; Available online: https://www.R-project.org/ (accessed on 2 April 2022).
23. Wang, Y.; Zhang, W.; Shi, T.; Xu, S.; Lu, B.; Qin, H.; Yu, L. Synergistic toxicity and physiological impact of thiamethoxam alone or in binary mixtures with three commonly used insecticides on honeybee. *Apidologie* **2020**, *51*, 395–405. [CrossRef]
24. Walker, E.K.; Brock, G.N.; Arvidson, R.S.; Johnson, R.M. Acute Toxicity of Fungicide–Insecticide–Adjuvant Combinations Applied to Almonds During Bloom on Adult Honey Bees. *Environ. Toxicol. Chem.* **2022**, *41*, 1042–1053. [CrossRef] [PubMed]
25. Belsky, J.; Biddinger, D.J.; Seiter, N.; Joshi, N.K. Various routes of formulated insecticide mixture whole-body acute contact toxicity to honey bees (Apis mellifera). *Environ. Chall.* **2022**, *6*, 100408. [CrossRef]
26. Strachecka, A.; Krauze, M.; Olszewski, K.; Borsuk, G.; Paleolog, J.; Merska, M.; Chobotow, J.; Bajda, M.; Grzywnowicz, K. Unexpectedly strong effect of caffeine on the vitality of western honeybees (*Apis mellifera*). *Biochem. Mosc.* **2014**, *79*, 1192–1201. [CrossRef] [PubMed]
27. Bajda, M.; Splitt, A.; Merska-Kazanowska, M. Effect of amphotericin B on the biochemical markers in the haemolymph of honey bees. *Med. Weter* **2014**, *70*, 766–769.
28. Strachecka, A.; Olszewski, K.; Paleolog, J.; Borsuk, G.; Bajda, M.; Krauze, M.; Merska, M.; Chobotow, J. Coenzyme q10 treatments influence the lifespan and key biochemical resistance systems in the honeybee, *Apis mellifera*. *Arch. Insect Biochem. Physiol.* **2014**, *86*, 165–179. [CrossRef]
29. Schulz, M.; Łoś, A.; Grzybek, M.; Ścibior, R.; Strachecka, A. Piperine as a new natural supplement with beneficial effects on the life-span and defence system of honeybees. *J. Agric. Sci.* **2019**, *157*, 140–149. [CrossRef]
30. Strachecka, A.J.; Olszewski, K.; Paleolog, J. Curcumin stimulates biochemical mechanisms of Apis mellifera resistance and extends the apian life-span. *J. Apic. Sci.* **2015**, *59*, 129–141. [CrossRef]
31. Strachecka, A.; Olszewski, K.; Paleolog, J. Varroa treatment with bromfenvinphos markedly suppresses honeybee biochemical defence levels. *Entomol. Exp. Appl.* **2016**, *160*, 57–71. [CrossRef]

MDPI AG
Grosspeteranlage 5
4052 Basel
Switzerland
Tel.: +41 61 683 77 34

Toxics Editorial Office
E-mail: toxics@mdpi.com
www.mdpi.com/journal/toxics

www.ingramcontent.com/pod-product-compliance
Lightning Source LLC
Chambersburg PA
CBHW062053220926
51850CB00010B/2751